The Art of Data Analysis

The Art of Data Analysis

How to Answer Almost Any
Question Using Basic Statistics

Kristin H. Jarman

Published by John Wiley & Sons, Inc. Hoboken, New Jersey
Published simultaneously in Canada

For general information on our other products and services or for technical support, please contact our Customer Care Department within the United States at (800) 762-2974, outside the United States at (317) 572-3993 or fax (317) 572-4002.

Wiley also publishes its books in a variety of electronic formats. Some content that appears in print may not be available in electronic formats. For more information about Wiley products, visit our web site at www.wiley.com.

Library of Congress Cataloging-in-Publication Data:

Jarman, Kristin H.
  The art of data analysis : how to answer almost any question using basic statistics / Kristin
  H. Jarman.
      pages cm
  Includes bibliographical references and index.
  ISBN 978-1-118-41131-5 (pbk)—ISBN 978-1-118-51492-4 (ePDF)—ISBN 978-1-118-51494-8 (eBook)—ISBN 978-1-118-51493-1 (ePub)—ISBN 978-1-118-51490-0 (eMOBI)   1. Mathematical statistics–Popular works.   I. Title.
   QA276.12.J35 2013
   519.5–dc23

                                                                    2013000381

10 9 8 7 6 5 4 3 2

To my father: The dream lives on.

Contents

Preface

I remember my first college statistics course. I studied hard and did the homework. I could calculate confidence intervals and perform hypothesis tests. I even earned a good grade. But the subject was so strange to me, I couldn't keep the different concepts straight. Populations, estimates, p-values, these things were nothing but a jumble of meaningless terms, and what little I learned vanished the moment I turned in the final exam.

Maybe I'm a masochist or maybe just determined, but I stuck with it. I took a second statistics course and then a third. It wasn't until I'd earned a Ph.D. in the field, worked on a number of real world problems, and made almost every mistake imaginable that I began to feel like I had a working grasp of statistics and its role in the data analysis process.

That's where this book comes in. It's driven by examples, not statistical concepts. Each chapter illustrates the application of basic statistics to a real dataset collected in the real world, far from the theorems and formulas and neatly contrived examples of the classroom. Hopefully, this book will provide you with the context you need to apply the basics of this slippery but oh-so-important subject to your own real-world problems.

Visit http://khjarman.com/ to contact the author, read more about the art of data analysis, and tell us your own statistical stories.

KRISTIN H. JARMAN

PART 1

The Basics

The Basics

CHAPTER 1

Statistics: The Life of the Party

As I sat in my favorite coffee shop, latte in hand, wondering how to introduce this book, my mind drifted to the conversations around me. At the table to my right sat a couple of college guys, decked out in sweatshirts touting a nearby university. They were arguing baseball, speculating which team was most likely to win the pennant and make it to the World Series. To my left sat three middle-aged women, speaking in quiet voices I had to strain to hear. They were talking about menopause, comparing their own experiences, trying to sort through the conflicting news about which, if any, treatments actually alleviate the symptoms. Behind me was the liveliest conversation of all. Two men were talking politics. Both men seemed to agree about who should win the next presidential election, but that didn't keep them from arguing. Many words were exchanged, but it came down to this. One of the men, citing a national poll, insisted his candidate was clearly going to be the winner. The other, citing yet another poll, claimed the outcome was anybody's guess.

Aside from my tendency to eavesdrop, there's a common theme to the three conversations. Whether they knew it or not, all of these people were talking statistics.

Most people run across statistics on a daily basis. In fact, in this age of instant information, it's hard to get away from them. Drug

The Art of Data Analysis: How to Answer Almost Any Question Using Basic Statistics, First Edition. Kristin H. Jarman.
© 2013 John Wiley & Sons, Inc. Published 2013 by John Wiley & Sons, Inc.

studies, stock market projections, sales trends, sports, education, crime reports: the list of places you'll find them goes on and on. Any time somebody takes a large amount of information and reduces it down to a few bullet points, that person is using statistics. And even if you never look at any raw data, when you use those bullet points to make conclusions or decisions, you're using statistics as well.

Being a statistician has never made me the life of the party. In fact, when I meet a new person, the reaction to my profession is almost universal. Here's how a typical conversation might go.

> ME: *How do you know John?*
>
> NEW FRIEND: *We work together at XYZ Corp. I'm in sales, and John's in marketing. How about you?*
>
> ME: *Oh, we've worked together on a few projects as well. So, are you married?*
>
> NEW FRIEND: *Yeah, sure. What projects?*
>
> ME: *The Michelson account, the Trends Survey, a few others. How 'bout kids? Do you have kids?*
>
> NEW FRIEND: *Hey, I worked on the Michelson account, too. Ran the sales campaign. But I don't remember you.*
>
> ME (glancing around the room): *I was just a consultant.*
>
> NEW FRIEND: *Hey, wait a minute. I do remember you. You told us we couldn't launch the product in June . . . something about summer and . . .*
>
> ME (shrugging): *Seasonal effects on customer demand. Yes, that's me. I'm the statistician.*
>
> *(There's a long, awkward pause. New Friend eyes me suspiciously.)*
>
> NEW FRIEND: *I always hated statistics.*

It may not be the life of the party, but when it comes to sorting through mounds of information, statistics is the belle of the ball. And it doesn't take a graduate degree in the subject to know how to use it. If you can apply a few basic statistical tools and a little practical knowledge to a problem, people think you're genius, and maybe even a little clairvoyant. These qualities may not draw crowds at the neighborhood mixer, but they do tend to result in big raises and big promotions.

Real-world statistics isn't only about calculating an average and a standard deviation. And it's not always a highly precise, exact science. Statistics involves gathering data and distilling large amounts of information down to a reasonable and accurate conclusion. Most statistical analyses begin not with a dataset, but with a question. What will be the impact of our new marketing campaign? Does this drug work? Who's most likely to win the next presidential election? Answering these questions takes more than a spreadsheet and a few formulas. It's a process: reducing the question down to a manageable size, collecting data, understanding what the data are telling you, and yes, eventually making some calculations. Often this process is as much an art as it is a science. And it is this art, the art of data analysis, that provides you with the tools you need to understand your data.

There are no proofs in pages that follow. Mathematical formulas are kept to a bare minimum. Instead, this book deals with the practical and very real-world problem of data analysis. Each chapter asks a question and illustrates how it might be answered using techniques taught in any introductory statistics course. Along the way, common issues come up, issues such as:

- How to turn a vaguely worded question into a scientific study
- How different types of statistical analyses are well-suited to different types of questions
- How a well-chosen plot can do most of the data analysis for you
- How to identify the limitations of a study
- What happens if your data aren't perfect
- How to avoid misleading or completely false conclusions

Every chapter is a case study, complete with a question, a data collection effort, and a statistical analysis. None of these case studies addresses society's fundamental problems (unless you think the lack of appreciation for superhero sidekicks is one of them). None of them will help you improve your company's sales (unless those sales are dependent on scientific proof that Bigfoot exists). And none of them will help you pick up women (especially not the one about gender stereotypes). On the other hand, all of them can be used as a template for your own data analysis, whether it be for a classroom project, a work-related problem, or a personal bet you just must win. And all of them illustrate how basic data analysis can be used to answer almost any question you can imagine.

The statistical techniques presented here can be found in most spreadsheet programs and basic data analysis software. I used Microsoft

Excel throughout, and in some cases, the Analysis Add-In pack was required. Here and there, a specific function is mentioned, but this isn't a book on statistics using Excel. There are plenty of good texts covering that topic. Some of the most popular, written by a man known as Mr. Spreadsheet, are listed in the Bibliography at the end of this chapter.

The outline of this book follows a typical introductory statistics course. Part One gives you the basic tools you need to ask a question and design a study to answer it. Part Two shows what you can do with a solid understanding of these basic tools. Each chapter is self-contained, but like a typical textbook, the concepts build on one another, and the analyses gradually become more sophisticated as the book progresses. If you're dying of curiosity and you've just got to find out when the zombie flu went viral, then go ahead and jump to Chapter 9. But if you can wait, I recommend you turn the page and read through the chapters in order.

I hope you enjoy reading these case studies as much as I enjoyed writing them.

BIBLIOGRAPHY

WALKENBACH, JOHN. 2007. *Excel 2007 Charts*. Wiley.
WALKENBACH, JOHN. 2010. *Excel 2010 Bible*. Wiley.
WALKENBACH, JOHN. 2010. *Excel 2010 Formulas*. Wiley.
WALKENBACH, JOHN. 2010. *John Walkenbach's Favorite Excel 2010 Tips and Tricks*. Wiley.

CHAPTER 2

Lions, and Tigers, and . . . Bigfoot? Oh, My: How Questionable Data Can Screw Up an Otherwise Perfectly Good Statistical Analysis

The mountain devil. Jacko. Sasquatch. Bigfoot. These are just a few names for a mysterious apelike creature rumored to be living in mountain forests around the United States. He's been a legend for generations. Ancient stone carvings of humanlike ape heads have been excavated in the Pacific Northwest (Eberhart 2001). Newspaper articles from the 1800s report "wild men" in such diverse geographic areas as Pennsylvania and California (Bord and Bord 2006). In the early 1900s, settlers and prospectors frequently reported seeing this creature in California, Washington, and Oregon (Bord and Bord 2006). There have been thousands of reported Bigfoot sightings in the last hundred years alone. And yet, no solid scientific proof of the creature exists.

The Art of Data Analysis: How to Answer Almost Any Question Using Basic Statistics,
First Edition. Kristin H. Jarman.
© 2013 John Wiley & Sons, Inc. Published 2013 by John Wiley & Sons, Inc.

Some of the eyewitnesses are con artists, to be sure. In 2008, for example, two gentlemen in Georgia threw slaughterhouse leftovers and a gorilla suit into a meat freezer, filmed the scene, and posted the video on YouTube, claiming they'd found the discovery of the century. The hoax only lasted a few days. Worldwide scrutiny soon got to the men, and they admitted it was all just a prank (CNN 2008).

Other sightings cannot be dismissed so easily. They come from seemingly reliable and trustworthy people, such as hunters, outdoorsmen, and soldiers, quiet residents who inhabit the very forests Bigfoot is reported to inhabit. Their reports are so vivid and so consistent, Bigfoot researchers have even compiled a detailed description of the creature, right down to the sounds he makes and his social behavior (Eberhart 2001).

The evidence doesn't end with eyewitness reports. Footprints, too large and square to be human, have been discovered, photographed, cast in plaster, and studied in detail. Hair samples of questionable origin have been collected. There's even a controversial film, shot in 1967 by Roger Patterson and Bob Gimlin. In this film, a large ape-man (or ape-woman, as some experts believe) walks through the forests of northern California. She's even so obliging as to glance at the camera while she passes.

With all this so-called evidence, Bigfoot researchers should have no trouble proving the existence of the creature. But this proof remains as elusive as the creature himself. Controversy over the Patterson film's authenticity rages on. DNA analysis of hair samples has, to date, been inconclusive. And even the best footprints somehow manage to look more fake than real. The most skeptical among us believe Bigfoot is pure myth, the subject of campfire stories and other tall tales. Others think the creature was real, a North American ape, perhaps, that once lived among early humans but long ago became extinct. Still others think the creature is alive and well, extremely shy, and living in remote areas across the United States.

Let's say, for argument's sake, I'm one of the believers. I'm convinced Bigfoot's real. Let's say, hypothetically speaking, I'm excited about the idea, so excited I just have to do something about it. I quit my job and head off in search of the creature. With visions of fame and fortune running through my head, I cash in my savings, say goodbye to my family, and drive away in my newly purchased vintage mini-bus.

As I leave the city limits, my thoughts turn to the task ahead. Bigfoot exists, there's no doubt about it. He's out there, waiting to be discovered. And who better than a statistician-turned-monster-hunter to discover him? I've got scientific objectivity, some newly acquired free time, and a really good GPS from Sergeant Bub's Army Surplus store. I've got only one problem. The United States is a big place, and no matter how much free time I have, I'm still only one person with a sleeping bag and a video camera. If I simply head to the nearest mountains and pitch my tent, the odds of me spotting a giant ape-man are about the same as the odds of me winning the lottery (and I don't play the lottery). No, I need to do better than put myself in the woods and hope for the best. But how will I do this? How will I ever prove Bigfoot exists?

GETTING GOOD DATA: WHY IT PAYS TO BE A CONTROL FREAK

It's too late to get my job back, and my husband isn't taking my calls, so it seems I have no choice but to continue my search. I decide I'm going to do it right. I may never find the proof I'm looking for, but I'll give it my best, most scientific effort. Whatever evidence I find will stand up to the scrutiny of my ex-boss, my family, and all those newspaper reporters who'll be pounding on my door, begging for interviews.

Having worked with data for nearly twenty years, I've learned that any conclusions you make are only as good as the data you use to make them. You may have all the statistical analysis tools in the world at your disposal, but without reliable data, they're useless. For example, suppose you're in the woods and you come across a large, oddly square-shaped footprint in the mud. Before you jump to conclusions and set up a press conference, you should check your data. Is there a bear nearby that might've made the footprint? How about a human? Is the footprint deep enough to have been made by a 700-pound primate? In the end, you may just find that your big discovery is really nothing more than a hole in the mud.

What I need is a study. The purpose of any study or experiment is to take a bunch of data and use those data to make conclusions about an entire group, or population. Experimental planning and design is the

1. Ask the question and determine whether a controlled experiment or an observational study is needed to answer it.
2. Specify the population and the effect you want to measure.
3. List all relevant factors. Use replication, random sampling, and blocking to prevent them from becoming confounding factors.
4. Make a data analysis plan. Check that it's consistent with the study and make adjustments as needed.

Figure 2.1.   Experimental planning and design.

process of planning a study, and it includes everything from deciding where your data will come from to how it will be analyzed. This process uses a handful of techniques to reduce the likelihood that your data will lead you to ambiguous, inaccurate, or even dead wrong conclusions. Taking the time to go through this process always pays off. It makes your data easier to analyze and your results easier to interpret.

The process of planning a study fits into the scientific method, a step-by-step approach to using measurements and observations to answer questions about the world around you (Wilson 2012). There are variations in how this process is presented. I've broken it down into four steps. These steps, detailed below, are summarized in Figure 2.1.

1. Ask the Question and Determine the Type of Study Needed to Answer It

A *study* is any data collection exercise. The purpose of any study is to answer a question. Is my company's marketing campaign working? Does this drug work? Does Bigfoot exist? Answering this question is the focus of the study. Throughout the planning process, lots of questions come up, and it can be easy to lose this focus, so I recommend writing down the original question and referring back to it often. This helps prevent you from adding unnecessary variables, following side paths, and otherwise becoming distracted.

Once the question has been clearly articulated, it's time to design a study to answer it. At one end of the spectrum, a study can be a *controlled experiment*, deliberate and structured, where researchers act like the ultimate control freaks, manipulating everything from the gender of their test subjects to the humidity in the room. Scientific studies, the kind run by men in white lab coats and safety goggles, are often controlled experiments. At the other end of the spectrum, an *observational study* is simply the process of watching something unfold without trying to impact the outcome in any way. When you say "yes" to that pop-up window asking you to participate in an online survey, you're probably joining an observational study. The researchers aren't trying to manipulate you; they just want to record how you feel about their product.

Controlled Experiments

Controlled experiments usually produce the best data and the strongest conclusions. In this type of study, all variables that could impact the outcome of the experiment are carefully controlled or measured. When researchers are able to run studies this way, by tweaking variables and observing the outcome, they can make statements about cause and effect. Cause and effect is the strongest type of conclusion, taking the form A causes B to happen. For example, about once a week, I prove that hot coffee burns my tongue. I know this because on those days I start sipping the moment the barista hands me my drink, my tongue screams in pain. On those days I wait ten minutes for my coffee to cool down, my tongue is fine. All the other important variables in this experiment are the same—me, the coffee shop, the kind of drink I order. The only things that change are the temperature of the coffee and the level of pain in my mouth.

Suppose, just for the sake of argument, I'm at Sergeant Bub's Army Surplus store, staring at a wall of cameras. There are wildlife cameras, motion activated cameras, infrared cameras, and any other type of camera that might appeal to outdoor enthusiasts, wildlife watchers, and the hopelessly paranoid. I'm looking for the perfect motion activated camera, one that will capture Bigfoot with Hollywood movie-like clarity. I know nothing about these devices, but there are two gentlemen standing next to me who seem like they do. I listen in, hoping for a little free expert advice.

"The Motion Sensor 3000 is better'n all the others," says one of the gentlemen, a man who looks like Santa Claus in camouflage. "It's high def. Also has the fastest response on the market."

"That piece of junk?" responds his friend, a dark, bony man dressed nearly head-to-toe in flannel. "Last one of them I had broke on me after only a couple days. Naw, what we want is the BearCam A110. It's the only camera worth buyin'."

The men go back and forth a while, arguing the merits of each camera, and this only adds to my confusion. Which model should I buy? I have limited resources and can't afford a camera that breaks. On the other hand, if I'm to capture Bigfoot on film, I need something state of the art, something that responds quickly and takes a good picture. I decide to design a study to pit these two cameras against one another. Because I'll be able to manipulate the conditions of my study and measure each camera's response, this will be a controlled experiment.

Observational Studies

Observational studies are those studies where the researcher cannot or does not manipulate any of the variables. He or she simply observes the outcome. It's difficult to assign cause and effect in an observational study. Why? When researchers are unable to measure the impact caused by tweaking a single variable, there's always the possibility another variable could be contributing to the result. That's why the findings of observational studies are often reported as associations, not cause and effect. For example, consider a study linking exercise to better health, where researchers have recorded the exercise habits of a thousand people, measured their health data, and found that people who work out frequently have lower cholesterol, lower blood pressure, and lower body fat. Because researchers do not manipulate any variables, this is an observational study. But we know there are many things besides exercise that affect a person's health: age, genetics, diet, and so on. These things are different for every person in the study. None of them can be controlled, and only a few, such as age and sex, can be observed. So whatever the result, it's impossible to conclude that exercise is the one and only cause of better health in this study. All the researchers can conclude is that the more active people tend to be more fit, meaning exercise is *associated with* but not necessarily the cause of better health.

There's an important difference between causation and association, and yet the two concepts are often confused. The news is full of conclusions made by well-meaning people who take an association and turn it into cause and effect, and the consequences can be significant. I refer you to *Freakonomics: A Rogue Economist Explores the Hidden Side of Everything* (Levitt and Dubner 2009) for some examples of the impact such confusion has had on our society over the past forty years.

Back to Sergeant Bub's Army Surplus store, where I'm standing in front of a wall of cameras, wondering which brand to buy (hypothetically speaking, of course). Listening to Santa Claus and Mr. Flannel argue over which is the best camera has given me an idea. Rather than running a controlled experiment and testing the different cameras myself, I might conduct a survey of Sergeant Bub's customers. On this survey I'd ask questions about the reliability, response time, and picture quality of the different cameras. I can't manipulate important variables such as the customers who stop by the store that day, which cameras those customers have used, and how they've used them. I can only take down opinions, all the while collecting data about who's buying what product and why. This study, a customer survey, is an observational study.

2. Specify Your Population and the Effect You Want to Measure

All studies draw data from a population. A *population* is the whole collection of people, places, or things under study. It's the group you want to analyze, to make conclusions about. You do this by collecting a *sample* of the population, in other words, a subset of the whole. Think of a controlled experiment in which I pit two different motion activated camera models against one another, one being the Motion Sensor 3000 and the other being the BearCam A110. The population consists of all Motion Sensor 3000 and BearCam A110 cameras. My sample includes only those cameras I'll get my hands on, the ones I'll test, the ones that will be representing the entire population.

Choosing and organizing a sample is a crucial part of the experimental design process. Statistically speaking, the best type of sample is called a *random sample*. A random sample is a subset of the entire population, chosen so each member is equally likely to be picked. For example, a random sample of Motion Sensor 3000 cameras might consist of ten such devices bought at ten randomly selected stores across town. Random sampling is the best way to guarantee you've chosen objectively, without personal preference or bias. Why is this important? Suppose Sergeant Bub has only two cameras for me to test. Both are from the back room, and both are customer returns. Because these cameras were returned to the store, I might expect one or both of them to be less than perfect, slightly flawed, or downright defective. If this is the case, then the performance of these cameras in my head-to-head comparison won't necessarily be typical of all Motion Sensor 3000 devices. In other words, my conclusions won't apply to the whole population.

Of course, this is the real world and pure random samples aren't always possible. Practical or ethical limitations often prevent us from this ideal. For example, to test a camera's sturdiness, I can't simply pull it from Sergeant Bub's shelf and throw it to the ground. Nor can I pick customers at random and force them to participate in my survey. In both cases, I need to get permission first. Because of these difficulties, we simply do the best we can, picking random samples whenever possible, and pointing out the limitations of our study when it isn't possible.

The *effect* is the outcome you want to observe. It should be something you can measure and analyze, and it should directly relate to the question in Step 1. In the head-to-head comparison of motion activated cameras, I want to pick a sturdy device that will be sure to capture any giant humanoids in its field of view. In this case, the effects for my study might include the time it takes for the device to respond to motion and the number of drops it takes before the device breaks. If, instead of comparing the cameras directly, I choose to do a customer survey, the effect might be a person's perceived reliability of each camera, as measured on a scale from one to five.

3. List All Relevant Factors and Decide What to Do with Them

Variables, or *factors*, include anything that can impact the results of an experiment, anything that can affect your effect, as it were. Factors fall into two basic categories: controllable and uncontrollable. *Controllable factors* are those variables you can manipulate in a study, those loved by scientists and control freaks alike. For example, I might want to test the response time of different motion activated cameras for different-sized objects to see which works best for a giant ape-man. The objects I put in front of the cameras while testing them are controllable factors. *Uncontrollable factors* are those variables you cannot manipulate. If I conduct my camera study outside, the weather is an uncontrollable factor. I can't change the cloud cover or the wind, but low sunlight and turbulent air movement could definitely have an impact on the performance of a motion-activated camera.

How do scientists deal with these different types of factors? Controllable factors are either systematically manipulated, so their impact can be measured, or they're held constant, so they don't contribute to changes in the effect being studied. Uncontrollable factors are observed whenever possible, so their impact can be taken into account during data analysis. In either case, it's important to consider all factors when planning a study. Any factor you don't account for can become a *confounding factor*. A confounding factor is any variable that confuses the conclusions of your study, or makes them ambiguous. For example, suppose I go to the vacant lot behind Sergeant Bub's and set up my camera test. On the first day, I test five Motion Sensor 3000 cameras, carefully measuring the response time of each. On the second day, I repeat the test for five BearCam A110 cameras. I've got my data, there's only one problem. The first day was bright and sunny. The second day was cloudy and windy. This means all of the Motion Sensor 3000 models were tested in one set of weather conditions, while all of the BearCam A110 cameras were tested in another. Weather, particularly poor light and wind, might impact a motion camera's response time, so whatever differences I observed could have been caused not by differences between the cameras themselves, but by the varying weather conditions. In other words, weather is a confounding factor in this study because it adds confusion to my results.

Confounding factors can really screw up an otherwise perfectly good statistical analysis. They lead to studies with no useful conclusions. They leave you standing in front of your boss, saying things like, "The upward trend in sunglasses sales could be due to our new marketing campaign, but it could also be due to the hot and sunny summer we've had. Since we didn't track sales last winter when it was cloudy, there's no way to tell." Fortunately, there are a variety of techniques that have been developed to minimize the chances this will happen.

Replication is the process of taking more than one observation or measurement. In the motion-activated camera experiment, this means more than one camera of each model should be tested. In the customer survey, it means opinions from more than one customer should be collected. Replication helps eliminate negative effects of uncontrollable factors, because it keeps us from getting fooled by a single, unusual outcome. As you'll see in future chapters, replication also adds certainty to the results a study produces.

In the same way choosing a random sample from a population eliminates biases in your results, *random sampling* eliminates biases due to how you collect your data. In random sampling, members of a sample are assigned a group or position at random. For example, to eliminate possible weather changes as a confounding factor in my camera comparison, I might test cameras in a random order, making sure to mix up the order in which the Motion Sensor 3000 and BearCam A110 are evaluated. Like replication, random sampling is an effective way to eliminate negative effects of uncontrollable factors.

Although it's a little more complicated than the first two techniques, *blocking* is a powerful way to eliminate confounding factors. Blocking is the process of dividing a sample into one or more similar groups, or blocks, so that samples in each block have certain factors in common. This technique is a great way to gain a little control over an experiment with lots of uncontrollable factors. For example, suppose I decide to test the different cameras under different lighting conditions to see how well they perform during the night and day. I might split them into two blocks, one to be tested at noon and the other to be tested at midnight. Each block should contain about the same number of both models. This way, each type of camera will be exposed to the same conditions throughout the course of the study, eliminating the light level as a potential confounding factor.

4. Make a Data Analysis Plan

It may seem strange to talk about analyzing data before it's even been collected, but this is the time to think about it. As you'll see in following chapters, different types of questions are well-suited to different types of statistical analyses. Each type of statistical analysis has its own requirements on the data needed to form a conclusion. By spending a little time visualizing how you will analyze the data and what form your results will take, you can go back through the first three steps of the experimental planning process to make sure you block, control, or observe all of the relevant variables in your study.

I like to start my data analysis plan by imagining a single chart or table that summarizes all of my findings, that one winning slide that impresses my boss and gets me a big promotion. I think through the data analysis process to get me to that magical slide. For example, hypothetically speaking, if I were planning a study to test two different motion activated cameras, I'd think about a single summary slide that would convey all my findings to my boss, a man who likes small words and very few bullet points. In this case, the winning slide might be a table comparing the Motion Sensor 3000 to the BearCam A110 in two environments: utter darkness and full sunlight. However, it wouldn't be enough to assume that it's completely dark at midnight and completely sunny at noon. A full moon or a rainstorm could change this. So, I'd take this opportunity to add one more variable to the data collection plan, the amount of light observed during the testing times.

FOOTPRINTS, FUR, AND A LIFE-CHANGING FILM

According to legend, Bigfoot is big, shaggy, and smells like a skunk that's been rolling in rotten meat. He looks like an ape but walks upright like a human. He likes to screech and whistle during the night. Occasionally, he throws rocks at passing humans. Finding such a creature ought to be a breeze, I figure, I just need to know exactly where to look.

I start narrowing down my search with step one of the experimental planning process, and I start by articulating my question. I stay away from philosophical questions like "Does the creature exist?" because unless I find the proof I'm looking for, there's no way to answer this question with any certainty. Instead, I limit myself to something concrete and measurable, something based on any suspicious evidence I might find, something like "Can humans and known animals be excluded as the source of any footprints, fur, or film footage I collect?" Since my search will take place out in a forest where I can't manipulate anything other than my own campfire, this will be an observational study.

Bigfoot's been spotted in mountain forests throughout the United States. If I had unlimited time and resources, I could select a random sample of mountain forests and search every one of them. However, I'm only one person and there are a lot of mountains in the country. So before I select my sample, I'll narrow my search with some background research. Fortunately, the Bigfoot Field Researchers Organization (BFRO) has done most of this research for me. The BFRO is a group of people devoted, in part, to investigating and documenting Bigfoot sightings around the country (Bigfoot Field Researchers Organization 2012). Every year, they receive hundreds of reports of Bigfoot encounters. They investigate each report, interview witnesses, and collect any available evidence. Those reports that pass a sniff test for hoaxes and pranks are posted on the organization's website, conveniently categorized by state and county.

According to bfro.net, the West Coast is the best place to conduct my search. As of this writing, California, Oregon, and Washington have seen over 1,000 unexplained, Bigfoot-like encounters since official record keeping began in 1995. Washington alone has had over half of these encounters. So, that's where I'm headed, to the land of rain, great coffee, and seven foot tall apelike creatures who like to lob boulders at tourists.

Washington has thirty-nine counties, some with no reported Bigfoot sightings, some with over fifty. I could use the number of sightings to pinpoint the location of my search, but aside from being a big state, Washington is also a divided state. Most of the population is concentrated in the three counties around the Seattle area. According to bfro. net, most of the Bigfoot-like encounters have occurred in Pierce County, an area that includes part of Seattle and Tacoma. I don't doubt all those eyewitnesses who swear the creature is living among the strip malls and forested neighborhoods surrounding the city, but if he were really there, it seems someone would've gotten proof by now. For my money, I'd rather search an area with an unusually large number of sightings compared to the population size. Skamania County, near Mount St. Helens volcano, is just such an area. With a little over 11,000 people (U.S. Census Bureau, Population Division, 2010 estimate) and more than 50 sightings since 1995 (bfro.net), this county sees five Bigfoot-like encounters for every thousand residents, more than any other area in the state.

Skamania County is nearly 1,700 square miles of small towns and wilderness. That's still a lot of ground to cover, but much less than the entire United States. For my study, I'll sample this area. Using my GPS, I'll choose wilderness locations at random and search them one by one until I find the creature or until my money runs out.

A seven-foot-tall shaggy beast who screeches and stinks up the forest ought to leave an abundance of evidence behind. My proof could include film; recordings of strange animal sounds; plaster casts of abnormally large, square footprints; and unidentifiable brown fur. These will be my effects, the variables I'm looking for. Fortunately, I've already bought all the equipment I need to gather the data: thermal cameras, motion activated cameras, and tape recorders to capture suspicious sights and sounds, plaster of Paris for casting footprints, and evidence collection jars for anything else I might find. I depleted my savings on as many of these things as I could afford, hoping to sprinkle the forest with replicate devices in order to increase my odds of success.

There are probably dozens of potential confounding factors in a study like this. I can think of two big ones: humans and hoaxes. First, there could be a human afoot in the forest, maybe a large, hairy lumberjack. If I'm not careful, I could confuse this guy's footprint or hair with Bigfoot's. To minimize the effect of this factor, human activity, I'll choose to limit my search to the most remote areas of Skamania County, places humans normally don't go. I'll also limit my study to a quiet time of year, early spring maybe, after the snowmobilers have gone but before the campers arrive, when all the animals are venturing away from their nests, caves, and dens to enjoy the spring weather.

I also need to think about a possible hoax. I'll be conducting my study in remote forests, where I won't likely run across any humans, much less ones pulling a prank, but if the locals learn of my expedition, one of them might decide to have a little fun with me, wandering through the forest in a giant ape-suit just to get me all worked up. To minimize the impact of this factor, what I call the "dupe-the-tourist factor," I'll keep my little study a secret, at least until I've uncovered the proof I need.

Finally, assuming Bigfoot really does exist out there, my very presence may impact my ability to find proof of the creature. After all, if Bigfoot came storming into my house and plunked himself down on my sofa, I'd probably find it difficult to go about my life as if nothing happened. Likewise, if I pitch a tent in the middle of Bigfoot territory and start poking around his home, the creature is likely to behave a little differently from how he normally would. So if I want to capture the creature on film, I need to plant the cameras and tape recorders in random locations, and leave. I'll return every few days to check up on my equipment and review the footage I've collected. If I find a suspicious video or an unexplainable screeching noise on my devices, I'll scour the immediate area for physical evidence: fur, footprints, scat, and so on.

The last step in the planning process, developing a data analysis plan, is pretty straightforward. For example, I imagine myself in front of the reporters, holding a vial of not-quite-human and not-quite-ape hair, announcing my discovery to the world. I see doubt on everyone's faces, and one reporter asks how I know this belongs to Bigfoot and not a bear. Of course, I'm convinced. I've reviewed the evidence with an objective, critical eye. Still, it's a good question, and one I won't be able to answer unless I have confirmation from an independent hair expert. So, I take this opportunity not to include another variable in my data collection process, but to set aside some of my remaining budget for expert forensic analysis of any evidence I find. This will add certainty to my conclusions.

One solid piece of evidence is all I need to restore my reputation and make amends with my family. I have a solid experimental plan, complete with a well-formulated question, a sampling scheme, and a method for eliminating potential confounding factors. I know where I need to go and what I need to do. All that remains is for me to cash in on the proof I know is out there, the proof I'm searching for, the proof I'm going to find.

Hypothetically speaking, of course.

BIBLIOGRAPHY

BIGFOOT FIELD RESEARCHERS ORGANIZATION. http://bfro.net/, accessed September 18, 2012.

BORD, JANET, COLIN BORD, and LOREN COLEMAN. 2006. *Bigfoot Casebook Updated: Sighting and Encounters from 1818 to 2004*. Pine Winds Press, Enumclaw, WA.

CNN. August 21, 2008. "Bigfoot Hoaxers Say It Was Just 'a Big Joke.'" http://edition.cnn.com/2008/US/08/21/bigfoot.hoax/.

EBERHART, GEORGE M. 2001. *Mysterious Creatures: A Guide to Cryptozoology*, vol. 1: *A–M*. ABC-CLIO.

eMATHZONE. "Basic Principles of Experimental Designs." http://www.emathzone.com/tutorials/basic-statistics/basic-principles-of-experimental-designs.html, accessed September 18, 2012.

LEVITT, STEVEN D., and STEPHEN J. DUBNER. 2009. *Freakonomics: A Rogue Economist Explores the Hidden Side of Everything*. Harper Collins.

NIST. "What Is Experimental Design?" http://www.itl.nist.gov/div898/handbook/pri/section1/pri11.htm, accessed September 18, 2012.

SCRUGGS, CATHERINE. "Basic Principles of Experimental Design and Data Analysis." http://www.ehow.com/info_8094232_basic-experimental-design-data-analysis.html, accessed September 18, 2012.

U.S. CENSUS BUREAU, POPULATION DIVISION. Release date April 2012. Table 1. Annual Estimates of the Resident Population for Counties of Washington: April 1, 2010, to July 1, 2011 (CO-EST2011-01-53), www.census.gov.

WILSON, JUDITH. "How to Plan and Design Experiments." http://www.ehow.com/how_8490451_plan-design-experiments.html, accessed September 18, 2012.

CHAPTER 3

Asteroid Belts and Spandex Cars: Using Descriptive Statistics to Answer Your Most Weighty Questions

Dear Mom,

I hope you're doing well. I haven't heard from you in over two weeks, and I worry you're still mad at me about the incident. I'd prefer to talk to you in person, but you don't seem to be getting my phone calls and you didn't answer the door when I visited. So, I must resort to explaining myself in an email.

First, when I dropped by your house with my new book project, I had no idea it would cause so much trouble. It's only a statistics text, after all. Yes, I probably should've called ahead, but in my defense, you asked me to bring a copy over some time. How was I to know you'd be hosting a party for the Shady Oaks Estates Ladies Club when I got there? And how was I to know you'd immediately start showing the book to all your guests before you even looked at it yourself?

The Art of Data Analysis: How to Answer Almost Any Question Using Basic Statistics, First Edition. Kristin H. Jarman.

Second, I agree "Your mama's so heavy . . ." jokes are rude. That's kind of the point. But please believe me, none of the insults in the book were meant to refer to you or any of your friends, especially not Marta.

Third, there's no need to wash out my mouth with soap. Yes, some of the jokes had foul language, rude references, and politically incorrect words. But I've removed the worst offenders and I hope you'll find the revised chapter much less objectionable.

Maybe you're right and insults like these have no place in a legitimate textbook. But you see, descriptive statistics are a lot like adjectives. An adjective, as you know, is a word that describes a person, place, or thing. A descriptive statistic is a number that describes a dataset. As you pointed out, "Your mama's so heavy . . ." jokes are full of colorful adjectives. What better way to illustrate descriptive statistics than to use them to answer the question, "How heavy is she?"

I sincerely apologize for the ruckus this chapter caused at your last party. I never meant to offend anyone. Please convey this apology to all your friends. Also, tell Marta the joke about the lady so heavy she deep fries her toothpaste was not a reference to her. Call me sometime.

Sincerely,
Your Loving Daughter

A recent Google search for "your mama" jokes came back with more than two million hits. Two million websites insulting "your mama." (And when I say "your mama," I don't mean your mom, the woman who gave birth to you and raised you to be the fine citizen you are today. I mean that other guy's mom. You know the one.) Even discounting duplicates, that's a lot of jokes, a lot of descriptions of this unfortunately-sized woman. Take these three jokes, for example:

Your mama's so heavy, she shows up on radar.
Your mama's so heavy, she's been zoned for commercial development.
Your mama's so heavy, her belt size is Equator.[1]

These three descriptions paint very different pictures, and if you were to use these descriptions to estimate the woman's size, you'd most likely come up with three very different answers. In other words, you'd have *variation* in your data. Virtually all real-world datasets have

---

1 Excerpts from Yo' Mama Is So. . . by Hugh Payne © 2007 used with the permission of Black Dog & Leventhal Publishers.

variation, or differences between observations. So how do you describe a dataset in the presence of variation? You use descriptive statistics. *Descriptive statistics* are numbers that summarize properties of a dataset. For example, suppose I had ten insults describing "your mama," a woman whom I'd never laid eyes on. I might take each of those insults and use it to calculate the weight and waistline of this woman. These calculated values would make up my dataset, and from them, I might be able to say the following:

"Your mama" is typically estimated to weigh 3,182 lbs.
Her belt size falls somewhere between Wide Load and Equator.
There's an 80% chance she drives a spandex car.

The typical weight. The range of her belt size. The likelihood she's big enough to require a car made of spandex. All of these values are descriptive statistics that tell you something about the woman whose size is in question.

Descriptive statistics are *estimates*, values calculated from a sample, values that approximate some property of the entire population. The average is a commonly used estimate. It's calculated from a sample of data and it approximates the typical value of a population. More on this descriptive statistic later. Life in the information age is full of descriptive statistics. Whenever a drug commercial warns its product might cause spontaneous bleeding, or a cable newscaster declares the economy is on life support, those statements are based on descriptive statistics. Whenever you search for a website, participate in a Facebook poll, or submit a customer review of your new cell phone, the information you provide gets combined with others' into a dataset, a dataset that's eventually summarized by someone—a search engine company, a friend, a marketing department. In short, wherever you've got data, you've got descriptive statistics, and they can be used to summarize virtually anything. The likelihood a person will search "funky chicken" on the Internet. The average number of whoopee cushions sold in stores last quarter. Or, as you'll see in this chapter, the typical belt size of a woman who fits the description of "your mama."

THEY MAMA

With over a half million web pages devoted to "your mama," gathering jokes is easy. However, it would take years for me to collect every insult in the world, especially considering the fact that new ones are constantly being created. Fortunately, I don't need every insult to describe this demographic, a group of ladies I'll call "they mama." I only need a sample, a subset of insults that represents the whole population of all "your mama" jokes in the world. As mentioned in Chapter 2, the most objective type of sample is a random sample, where every insult has the same likelihood of being included. Random sampling is especially important for a case study such as this, where my own personal joke preferences could cause me to pick certain insults over others, resulting in a skewed picture of "they mama." To collect my sample, I went to two of the biggest user-contributed joke websites, *Aha! Jokes* and *Yo' Mama Jokes Galore*, and chose two hundred "your mama so heavy" jokes at random.

Like the "your mama" who weighs herself on the Richter scale, the difference between ladies in "they mama" is huge. For example, the woman who needs to grease the door when she enters the house is petite compared to the one who's been named Miss Arizona . . . Class Battleship. Miss Arizona Class Battleship is tiny compared to the woman who influences the tides. With so much variation, I need descriptive statistics to give me an idea of just how big the ladies in this demographic are. But first, I need to turn these colorful yet vague insults into data.

QUALITATIVE VERSUS QUANTITATIVE DATA

All data fall into two categories: qualitative and quantitative. *Qualitative observations or data* are typically categories, groups or characteristics. Hair color and favorite foods are examples of qualitative observations. *Quantitative observations or data* are numerical values. Weight and belt size are examples of quantitative observations.

These two types of data are generally treated differently. The reason? Plain and simple arithmetic. Qualitative observations cannot be sorted into a numerical order. For example, suppose you're analyzing the hair color of a group of ladies. You might take each lady and categorize her into one of a few groups: blonde, brown, red, black, and gray. The color brown isn't larger or smaller than red. It's just different. And without a mathematical relationship between observations, we're somewhat limited in our ability to mathematically summarize qualitative data. Quantitative observations, on the other hand, are meaningful numerical values and so they can be sorted. If you're weighing the ladies in "they mama," for example, 4,500 lbs is heavier than 4,400 lbs, which is heavier than 4,350 lbs, and so on. This mathematical ordering allows us to use the full arsenal of arithmetic, algebra, and even calculus to summarize quantitative data. I'll describe "they mama" both ways, starting with the simpler of the two approaches.

Qualitative Analysis

After reading over my list of insults gathered from the Internet, I began to notice a pattern. While the details of the different jokes vary, many of them compare "your mama's" size to the same small number of objects—cars, whales, buildings, and so on. These comparisons give me a convenient way to categorize the insults, turning a bunch of one liners into qualitative data. After poring over the list a few more times, I settled on seven categories of ladies: (1) *Unimpressive*, (2) *Large Mammals*, (3) *Planes, Trains, and Automobiles*, (4) *Buildings*, (5) *Geological and Geographical Phenomena*, (6) *Astronomical Objects*, and (7) *Who Knows?* The first category, *Unimpressive,* includes women whose size, while large, is nothing particularly special. The sizes of the ladies in groups two through six are implied by the category labels and should be obvious. The last category, *Who Knows?*, is a catch-all group of insults without any obvious, concrete reference. Many of these insults have something to do with the woman's eating habits or clothing challenges. Examples of jokes falling into each category are listed in Figure 3.1.

You might wonder why I didn't make planes a separate group from trains and automobiles, or why I didn't separate the women who cause geological phenomena from the women who take up large geographic areas. The answer is simple. I made a judgment call. I categorized the ladies this way because it makes sense to me. If you prefer a different grouping, I urge you to find your own jokes and repeat the following analysis for yourself.

Once all the ladies in "they mama" have been categorized, I have a set of qualitative observations to work with. Like all qualitative data, these have no clear mathematical ordering, and so they cannot be analyzed by any method that arithmetically compares different observations. So, I'll do what people usually do with data like these. I'll start with something called a frequency distribution.

Category	Example
Unimpressive	...when she steps on the scale, it says, "To be continued...".
Large Mammals	...she got baptized at SeaWorld.
Planes, Trains, and Automobiles	...she shows up on radar.
Buildings	...she's been zoned for commercial development.
Geographical and Geological Phenomena	...when she went to the beach, she caused a tsunami.
Astronomical Objects	...she wears an asteroid belt.
Who Knows?	...she deep fries her toothpaste.

Figure 3.1.   "Your mama's so heavy . . ."

The frequency distribution is a common way to summarize a set of observations. For qualitative data, the *frequency distribution* is just a list of counts—the number of observations falling into each category. Figure 3.2 lists the frequency distribution for "they mama." In the figure, the relative frequency is also included. The *relative frequency* represents the fraction (or alternatively, percentage) of all the observations falling into each category. This fraction is the ratio of the number of counts in each category to the total number of observations. The percentage is just the fraction multiplied by 100.

The frequency distribution in Figure 3.2 shows the relative popularity of different types of insults. For example, the smallest women in "they mama" are *Unimpressive*. Twenty-seven, or 14 percent, of the ladies fall into this group. On the other end of the body mass spectrum sits the *Astronomical Objects* category. This category may contain the largest women, but with only 4% of the insults, it's the least popular. The categories *Large Mammals* to *Astronomical Objects* include those women who are both impossibly large and compared to something concrete. Adding up the frequencies of these categories tells us these ladies make up 52% of all the jokes.

Category	Counts (Number of Insults)	Relative Frequency
Unimpressive	27	0.14
Large Mammals	8	0.04
Planes, Trains, and Automobiles	26	0.13
Buildings	21	0.11
Geographical and Geological Phenomena	41	0.21
Astronomical Objects	7	0.04
Who Knows?	70	0.35
Total	200	1.0

Figure 3.2. Frequency distribution of "they mama."

With relatively few observations and relatively few categories, tables like Figure 3.2 do a fair job of illustrating the frequency distribution. However, nothing compares to a good graph. Bar charts are particularly useful for frequency distributions. A bar chart displays each category as a box whose height represents the number of observations in that category. Figure 3.3 shows a bar chart of the "your mama" insults, plotted as the *relative frequency* in terms of percentage.

A quick glance at Figure 3.3 tells us a lot about these data. The *mode*, the most popular category in "they mama," is *Who Knows?* This category contains almost twice as many as any other. *Astronomical Objects*, with a relative frequency ten times lower than *Who Knows?*, is the least popular. The categories *Unimpressive*; *Planes, Trains, and Automobiles*; *Buildings*; and *Geographical and Geological Phenomena* are comparable, each having a relative frequency between 10 and 20%.

Aside from the mode, which tells you what category is most typical in your dataset, relative frequencies are the most common descriptive statistics used in the analysis of qualitative data. For example, the batting average of a baseball player is a relative frequency, the number of base hits divided by the total number of times at bat. The results of

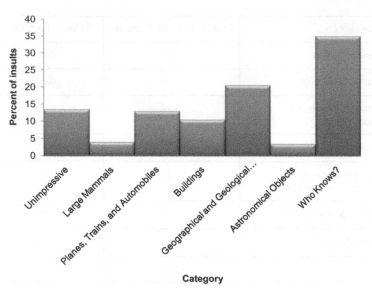

Figure 3.3.   "They mama" frequency distribution: how heavy is she?

political polls are typically relative frequencies, expressed as a percentage of people who approve of a particular candidate. Relative frequencies are particularly useful for calculating probabilities. This topic will be discussed more in the following chapter and throughout the rest of the book.

Quantitative Analysis

The frequency distribution in Figure 3.3 provides us with a good general summary of "they mama," but it doesn't really answer the question at hand. In order to determine exactly how heavy "your mama" is, I need more precise information, measurements that indicate her belt size or her weight. In other words, I need quantitative data. The kind of numerical values that'll only be possible with a little creative calculation.

For example, imagine a woman who's so large she drives a spandex car. What kind of car is it? Is it a spandex Corolla or a spandex Suburban? The answer makes a difference. According to cars.com, a Toyota Corolla is about 69 inches wide, where a Chevy Suburban measures in at roughly 79 inches wide. That's a difference of 10 inches in width, which translates to a 30-inch difference in the belt size of a woman who fills such a car.

Many of my insults are like this one, leaving much room for variation in the measurement. I could go through the jokes one by one, estimating a range of sizes for each and folding them together into one grand dataset. However, this would be a tedious process with many details to explain. So, rather than becoming distracted by creative calculations, I'll pick one insult and conduct a quantitative analysis on that. Here it is: "Your mama's so heavy she has her own zip code." Let's say the U.S. Postal Service assigns zip codes to ladies based on the size of their waistline. In other words, assume "your mama's" waistline takes up an area roughly the size of a typical zip code region. Exactly how big is this area?

The U.S. Census Bureau spends millions of dollars every year keeping track of the U.S. population—demographics, number of houses, income, number of children, and yes, the land area of every zip code in the country. I'm pretty sure the government never expected the data to be used to measure "your mama's" waistline, but it makes these data publicly available in any case. I should mention there's something special about these zip code areas. Where most datasets are merely samples, or subsets of the entire population, these data are complete, including every zip code in the country. In other words, I don't just have a sample here. I have the entire population.

As of this writing, the 2010 data are not yet available, but the 2000 Census zip code tabulation areas (ZCTAs) can be downloaded from www.census.gov. There are a little over 32,000 values in this dataset, each one representing the land area, in square miles, of a U.S. zip code. Unless you're squeamish about lots of values, I encourage you to download these data and repeat some of the following analysis yourself.

THE DESCRIPTIVE POWER OF STATISTICS

Imagine we're outside on a sunny day. You're blindfolded and I'm trying to describe a cloud in the sky. I might tell you where it is, whether high in the sky or near the horizon. I might tell you how big the cloud is, whether it takes up one tiny quadrant or looms over the entire landscape. I also might describe its texture, where it's thick and dense and where it's light and transparent. These three characteristics, location, size, and texture, would help you form a more detailed picture of the cloud in your mind. These same characteristics are the ones most commonly used to describe quantitative observations, a data cloud, if you will. In statistics jargon, the location is often called center location, the size is often referred to as variation, and the texture is often represented with the frequency distribution. I'll start by describing the last of these characteristics first.

Frequency Distributions and Histograms

The frequency distribution measures the relative popularity of every category in a qualitative dataset. For a quantitative dataset, it measures the texture of the data cloud, where the cloud is thick and dense with many observations, and where it is thin with few or none. Constructing the frequency distribution for quantitative data is a lot like the process for qualitative data, where observations in each category are tabulated and then plotted using a bar chart. In fact, for *discrete* quantitative data—observations that can only take on distinct values like integers from one to ten—the process is exactly the same. You simply count up the number of ones, twos, threes, and so on, and then list or graph the results. The differences arise when you have *continuous* quantitative data—values that take on all possible numbers in some specified range. Why? With continuous values, you can spend your entire life counting observations at any given value, but there will always be another value to count. For example, if your data can take on any value between 50 and 60 and you count the number of 56s and 57s, you still have 56.1, 56.2, 56.5, and lots of others. If you then count the number of 56.1s, 56.2s, and so on, you still have 56.15, 56.23, 57.28, and lots of others. You can stretch the decimal places all you want and there will still be more numbers to count. Infinitely many of them, in fact. Try plotting a bar graph with infinitely many categories on the x-axis. Go ahead. I'll wait.

For quantitative data, the frequency distribution is typically calculated by splitting the observations into discrete bins, or ranges of values, and counting the number of observations falling into each bin. For example, if you have observations between 50 and 60, you might construct ten bins, one for observations falling between 50 and 51, one for observations falling between 51 and 52, and so on. By binning the data in this way, you can tabulate frequencies and relative frequencies for each bin and then list or graph the results.

I've said this before, but it's worth repeating. The best way look at a large amount of data is with a graph. A graph helps you visualize your observations and it can be incredibly helpful in identifying *outliers*, extreme or unusual values that can impact your results. You've already seen how a bar graph can be used to visualize the frequency distribution for qualitative data. The same type of graph can be used to visualize the frequency distribution for quantitative data. This type of graph is so popular, it has its own name: a *histogram*.

When plotting a histogram, deciding how to bin the data is a bit of an art, but some guidelines are available. Plotting programs such as Excel divide the range into $\sqrt{N}$ equal-size bins (Microsoft Corporation 2012), but the best choice often depends more on what makes the most sense for your specific problem than what any spreadsheet software recommends. In general, bins are calculated from the *maximum*, the largest data value, and the *minimum*, the smallest data value, by dividing this range into equal length intervals. The number of bins you use can dramatically change the shape of the histogram, especially when your sample is small and you have only a few observations per bin. I recommend experimenting with different bin sizes. It'll help you decide if the shape you're seeing reflects the true nature of your data or simply your choice of bin widths.

For the zip code tabulation areas, the minimum area is 0.0019 square miles, about the size of a large building, and the maximum is 5,497 square miles, about the size of Connecticut. With just over 32,000 values, the square root criterion suggests 179 bins for this histogram, each with a width of about 31 square miles. However, this choice of bin number produces a fairly useless histogram, with over 90% of all zip code areas falling into the first two bins. Such a graph, with only two visible bars and no hints as to the shape of the distribution, doesn't really teach us anything. Instead, I'll increase the number of bins to 1,500. This breaks up the first few bins, making it easier to see what happens at the low end of the zip code areas. This histogram is plotted in Figure 3.4.

If you've been in your statistics class for more than a few weeks, you've probably heard about the bell-shaped frequency distribution. Illustrated in Figure 3.5, the bell-shaped distribution is the poster child

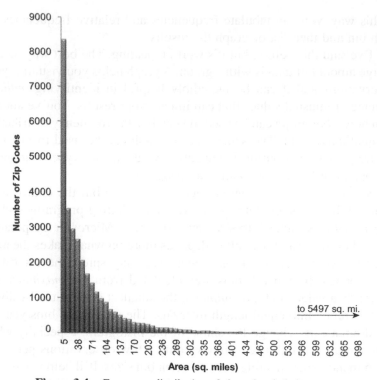

Figure 3.4. Frequency distribution of zip code tabulation areas.

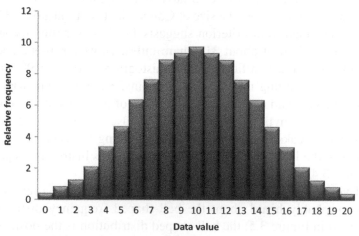

Figure 3.5. An illustration of the bell-shaped frequency distribution.

of all frequency distributions, with its central peak and gently sloping, symmetric sides. Many datasets have a bell-shaped distribution. The zip code dataset isn't one of them. There are a large number of very small zip code areas, less than about 16 square miles. Beyond that, the relative frequencies taper off slowly, from 16 all the way out to 5497 square miles (although the histogram has been cut off at 700 square miles so you can see details of the smaller zip code areas). It's like a cloud that's thick and dense on one end, and thin at the other. Because the histogram looks like it's been stretched to the right, this frequency distribution shape is called *right-skewed*. Right-skewed distributions are fairly common when it comes to measuring counts, distances, and areas. There's also a *left-skewed distribution*, whose histogram looks like it's been stretched to the left, but those are less common.

At this point, you might be asking yourself why you should care about the shape of the histogram. After all, we're only trying to estimate the typical belt size of a woman who has her own zip code. Does it really matter whether the data are symmetric, left-skewed, or right-skewed? As you'll see in the following section, it does.

Central Location

The term *central location* refers to the center of a data cloud, in other words, the spot around which all the data values are clustered. Two descriptive statistics are commonly used to measure central location: the sample mean and the median. Of those two, the sample mean, or the average, is most common.

Sample Mean (Average)

Even if you've never had a single teacher utter the word "statistics" in class, you've probably run across the average. Calculated by adding all the data values together and dividing by the number you have, the average pinpoints the arithmetic center of a dataset.

For measurement values x_1, x_2, . . . , x_N, the average is

$$\bar{x} = \frac{x_1 + x_2 + \cdots + x_N}{N}.$$

For example, the average of the four numbers 4, 5, 5, and 6 is $(4 + 5 + 5 + 6)/4 = 5$.

Median

The **median** is another common way to measure the central location of a data cloud. However, unlike the average, the median isn't calculated from an arithmetic formula. Rather, it's the midpoint, or middle measurement value. To calculate the median, sort your list of numbers from smallest to largest. If there's an odd number of values in the list, pick the one in the middle. If there's an even number, then average the two middle values. For example, the median of 4, 5, 5, and 6 is the average of the two middle numbers, 5 and 5, which is, of course, 5.

If a frequency distribution is *symmetric*, meaning the two halves of the histogram are mirror images of one another, the large values will balance the small values both arithmetically and in terms of the middle position. In a case like this, both the average and median will lie in the middle of the histogram, very close to one another. The sample mean and median of the bell-shaped data shown in Figure 3.5, for example, are both ten, the center position on the histogram. On the other hand, if the frequency distribution isn't symmetric, all bets are off. Extreme values and nonsymmetric bumps in a dataset can cause these two descriptive statistics to be quite different. The sample mean of the zip code areas is 85.8 square miles. The median, 37.7 square miles, is less than half that value.

Why should the sample mean and median be so different? These two statistics both measure the center of a data cloud, but they do it in two very different ways. Think of the numbers 4, 5, and 6, for example. The average of these three values is $(4 + 5 + 6)/3 = 5$. The median, the middle number, is also five. Now stretch the six to a nine, giving the values 4, 5, and 9. The average of these numbers has increased to $(4 + 5 + 9)/3 = 6$, but the median is still five. Stretch the 9 to a 12 and the average increases to 7. The median? Still 5. The average takes into account all the values in the dataset, even the very large or small ones. The median chooses the middle value, regardless of what's happening at the extremes. In other words, the mean is impacted by big or small values, while the median isn't nearly as much.

According to the sample mean and the median of the zip code tabulation areas, a typical "your mama's" waistline consumes anywhere from about 38 square miles (the median) to about 86 square miles (the sample mean). In other words, she takes up more space than Newark, New Jersey, and less than Amarillo, Texas. Which one is a better indicator of the center location of this data cloud? As you'll see in coming chapters, the sample mean is more common and, in many ways, easier to work with when doing data analysis. However, for purely descriptive purposes, it's a good idea to calculate both, because comparing the mean and median can tell you a lot about your data. If the two values agree with one another, chances are good you're looking at a nicely formed, symmetric frequency distribution. When they don't, you probably want to dive a little deeper into the data. Plot a histogram. Look at the extreme values. You might just have an oddly shaped or skewed distribution, something that's useful to know before you make any conclusions.

Variation

Variation refers to the size of a data cloud. Understanding variation is one of the most important parts of any statistical analysis. Why? Because people rely on statistics to make weighty decisions, and variation has a big impact on everything from the simplest data summary to the most sophisticated nonlinear analysis. Just think about the sample mean and median of the zip code data. These two descriptive statistics both measure central location, and so it seems they should agree with one another. But they don't. And this is because variation gives rise to a right-skewed frequency distribution in the data.

Variation pops up any time there's more than one person, place, thing, or measurement in a group. It doesn't matter if you're counting the number of molecules in a test tube or the number of defective parts coming off an assembly line. Variation just happens. Like central location, there are several ways to measure variation. I'll introduce three of the most common.

Range

The *range* is the simplest descriptive statistic for variation. It's the largest value, the maximum, minus the smallest value, the minimum. In other words, the range is the span of your data cloud. For example, the range of zip code areas is the difference between the largest and smallest values, or 5497.000 − 0.002 = 5496.998 square miles. That's the total variation of waistline sizes in "they mama." Simple, right?

Unfortunately, simple is not always better. Because it only takes into account the largest and smallest measurement values, the range is generally easy to calculate and easy to understand. But it ignores the bulk of the observations, the ones in the middle, and so it can also be terribly misleading. For example, suppose you have five observations: 5, 5, 5, 5, and 20. The average of these observations is 8. The range is 15. Without looking at every data value, you might be misled to think the bulk of observations are around 8, with values ranging from about zero to 15. But the actual frequency distribution looks nothing like this. In fact, all of the values are very tightly clustered at 5, with only a single extreme value, an **outlier**, at 20. Just like it impacts the average, this outlier impacts the range. Without this one value, the average would drop to 5, and the range would fall to zero.

The range is notoriously impacted by skewed distributions and outliers. This can be a good thing when you're doing an analysis of the extremes, but most descriptive summaries are concerned with the majority, not the unusual. In the analysis of zip code areas, the range may be nearly 5,500 square miles, but it's clear from Figure 3.4 the

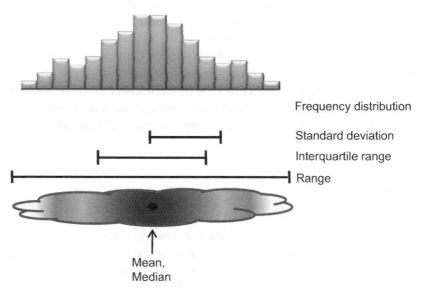

Figure 3.6. Describing the central location, variation, and texture of a typical data cloud.

vast majority of values are smaller than 400 square miles. This is where most of the data cloud lies.

Interquartile Range

Percentiles are sometimes used to understand a dataset. A *percentile* is the point below which some specified percentage of the observations fall. Think of your dataset as a list of values, sorted from smallest to largest. The 10th percentile is the value that's one-tenth of the way down the list. Ten percent of the values are smaller than the 10th percentile. Ninety percent of them are larger. The 50th percentile is the median, the value halfway down the list. *Quartiles* refer to the 25th, 50th, and 75th percentiles, those values 25%, 50%, and 75% of the way down the list.

Consider the list of data values: 2, 3, 4, 4, 5, 6, 6, 7, 7, 8. This list has ten numbers. The tenth percentile is the value one-tenth of the way down the sorted list. That's the first number on the list, or 2. The 30th percentile is 30% of the way down the list, the third value, which is 4. The 50th percentile, the median, is the fifth number, which is 5. The 75th percentile is 6.5.

The *interquartile range (IQR)* measures the spread of the middle half of data values. This statistic is calculated as the difference between the 75th and 25th percentile. For example, the list 2, 3, 4, 5, 5, 6, 6, 6, 7, 8, 9, 10 has twelve numbers. The 75th percentile is 7 and the 25th percentile is 4. The IQR is $7 - 4 = 3$. This means the middle half of the data values fall within three of one another. Because the IQR uses data values well inside the extremes, this statistic is always less than the range. It also tends to be less sensitive to outliers. For example, the total range of zip code areas is nearly 5,500 square miles. The IQR is $90.4 - 9.0 = 81.4$ square miles. This means the middle half of the values fall within 81.4 square miles of one another. The other half of the values are responsible for the rest of the total variation, all 5,419 square miles of it.

Standard Deviation and Variance

The *standard deviation* and its alternative form, the *variance*, are the most common measures of variation. Like the average, the standard deviation is an arithmetic value that uses all of the observations in its calculation. Here's the formula.

For measurement values $x_1, x_2, \ldots, x_N$, with average $\bar{x}$, the standard deviation is

$$s = \sqrt{\frac{(x_1 - \bar{x})^2 + (x_2 - \bar{x})^2 + \cdots + (x_N - \bar{x})^2}{N - 1}}.$$

The variance is the standard deviation squared, in other words, s^2.

For the list of values 4, 5, 5, and 6, with an average of 5, the variance is

$$s^2 = \frac{(4-5)^2 + (5-5)^2 + (5-5)^2 + (6-5)^2}{3} = 0.66$$

and the standard deviation is

$$s = \sqrt{\frac{(4-5)^2 + (5-5)^2 + (5-5)^2 + (6-5)^2}{3}} = 0.81.$$

The standard deviation doesn't measure the size of a data cloud, at least not directly. Rather, it measures the average deviation of your data around the sample mean. The standard deviation is always smaller than the range. And for bell-shaped data, it's typically a little smaller than the IQR as well. For the bell-shaped data in Figure 3.5, for example, the total range is 20, the IQR is 5.5, and the standard deviation is 3.9. For skewed data, the standard deviation is still smaller than the range, but it can be smaller or larger than the IQR, depending on how stretched the frequency distribution is. For example, the total range of the zip code areas is 5,500 square miles, and the IQR is 81. The standard deviation is 192 square miles. That's over twice the interquartile range, a strong indication of just how skewed these data are. Figure 3.7, while not to scale, illustrates the relationship between these three measures of variation as well as to the other descriptive statistics introduced in this chapter.

All of these measures of variation can be useful when analyzing quantitative data; however, the standard deviation plays a particularly important role in statistics. It may not be as straightforward as the range or the IQR, and it may be sensitive to outliers, but it's firmly rooted in the mathematical foundation of uncertainty, and, frankly, statisticians

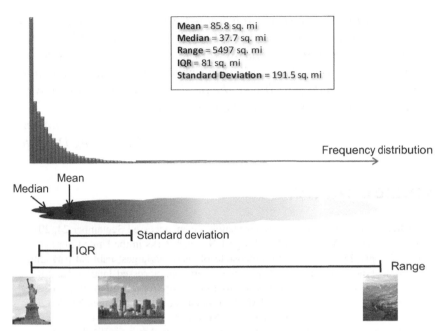

Figure 3.7.   Describing "your mama": how big is she?

love it. As you'll see in following chapters, the standard deviation leads to many useful statements about the size of a data cloud, helping you to calculate the margin of error in a set of observations, compare two or more datasets together, and determine if a trend you're seeing is real or coincidental.

HOW HEAVY IS SHE?

Dear Mom,
It's been two months now, and still I haven't heard from you. I hope it's because you've been busy with your duties as president of the Shady Oaks Ladies Club and not because you're still mad at me.

Along those lines, I have some good news. I've finished my chapter on descriptive statistics and guess what? There's no reason for your friends to be upset. My qualitative analysis showed that over half of the ladies in the "your mama" demographic are impossibly heavy, and by that I mean large mammal to planetary object heavy. On top of that, my quantitative analysis, though limited, shows that a typical "your mama" is so heavy, she could sell shade to the whole Shady Oaks Estates subdivision, all five phases of it! In other words, there's no real resemblance between these ladies and your friends. That means none of your friends could possibly be the inspiration for this chapter. None of them have any reason to be offended. Not even Marta.

Sincerely,
Your Loving Daughter

BIBLIOGRAPHY

Aha! Jokes. "Yo' Mama Jokes." www.ahajokes.com, accessed September 24, 2012.
BRINEY, AMANDA. May 29, 2011. "Largest National Parks in the United States." http://geography.about.com/od/unitedstatesofamerica/a/largest-national-parks.htm.
Cars.com. "Find New and Used Cars." www.cars.com, accessed January 12, 2012.
MICROSOFT CORPORATION. "Explore Histograms." http://office.microsoft.com/en-us/excel-help/explore-histograms-HA001110948.aspx, accessed September 24, 2012.
PAYNE, HUGH. 2007. *Yo' Mama Is So. . . .* New York: Black Dog & Leventhal.
UNITED STATES CENSUS BUREAU. Zip Code Tabulation Areas (ZCTAs™). http://www.census.gov/geo/ZCTA/zcta.html, accessed October 15, 2011.
"Yo Mama Jokes Galore!" http://www.yomamajokesgalore.com/, accessed September 24, 2012.

CHAPTER 4

Baseball, Game Shows, and Sidekicks: Who Cares about Probability Anyway?

Welcome to Gotham City, the metropolis that has it all: wealth, culture, high society, and the strangest assortment of arch-criminals ever seen. Fortunately for its residents, Gotham City also has Batman, a superhero who sports a cape, tights, and a bat cowl that covers his face, keeping his identity a secret. Whenever the likes of the Joker, the Riddler, or Catwoman are in town, the caped crusader is ready for action.

Batman's story began in May 1939 when the character, created by Bill Finger and Bob Kane, first fought crime in Detective Comics #27 (DC Comics 2004). The superhero became an instant hit, earning his own title within a year of his first appearance. The comic initially ran quarterly, then biweekly, then monthly. As the story developed, Batman took on several different personas, from darkly serious to campy and back again. After seventy years, he's still fighting crime, with ongoing DC Comics publications, graphic novels, movies, cartoons, television shows, and video games.

By day, Batman is the famous rich man and philanthropist Bruce Wayne. He inherited his fortune from his parents, who were killed in a hold up while walking home from the theater. Witnessing his parents'

The Art of Data Analysis: How to Answer Almost Any Question Using Basic Statistics, First Edition. Kristin H. Jarman.
© 2013 John Wiley & Sons, Inc. Published 2013 by John Wiley & Sons, Inc.

murder scarred young Bruce Wayne. But rather than choosing therapy, he opted for a secret identity and a life of vengeance. Armed with a genius intellect, the athleticism of an Olympian, and high tech gadgets developed in a top secret underground laboratory known as the Batcave, Bruce Wayne now takes out his revenge on any criminal foolish enough to target the citizens of Gotham City.

In the early years, the Golden Age as it is sometimes called, Batman was usually accompanied by his sidekick, Robin. First appearing in 1940, Robin fought by Batman's side, helping defeat such criminal masterminds as the Joker, the Penguin, Catwoman, and the Riddler. By day, Robin was Dick Grayson, an orphaned circus acrobat whose parents, much like Bruce Wayne's, were murdered by thugs. Batman took in Dick Grayson and raised him like a son while he initiated the boy into the life of a superhero. The Boy Wonder stayed with Batman for nearly 30 years, until 1969, when he grew up, went off to college, and became a superhero in his own right.

Batman is to Gotham City what statistics is to data analysis—a superhero ready to jump into the fray and solve any problem, no matter how small. The Joker holding the mayor for ransom? Catwoman robbing Gotham City Natural History Museum of its solid gold cat statues? The boss on your back to make those sales projections? No problem. Batman and statistics are there to help.

If statistics is the Batman of data analysis, then probability is his sidekick. While Batman's in the trenches, killing villains, saving children, and taking a beating, Robin stands right beside him, ready with a bottle of energy water and a couple ibuprofen as needed. Likewise, while statistics is out there, searching for truth among reams of data, probability stands nearby, ready to supply the mathematical support that fuels statistics through even the toughest data analyses.

Robin and probability are hard-working, loyal friends, but they come with baggage. Hard as he tries, Robin often ends up in a criminal's trap and Batman must save him. Probability can be difficult to understand, and even more difficult to apply to any sort of real world problem. This begs the question. If Robin and probability are merely sidekicks, do we really need them? After all, Batman carries a full sized batshield, a batarang, and a Batmobile remote transmitter in his utility belt. Surely, Robin can't offer much more than that. Likewise statistics, the science of making conclusions from data, does all the heavy lifting when it comes to analyzing data. Probability may be its mathematical foundation, but is it really needed when you're calculating an average and a standard deviation?

Are Robin and probability merely distractions, or do they offer something of value? Who cares about sidekicks, anyway?

MEASURING THE THREE Bs

According to my unofficial and completely unscientific research, a sidekick should be loyal. You don't need statistics to know Robin is loyal. Between 1940 and 1969 he was there, time and again, fighting crime and taking his lumps right alongside Batman. Even after he grew up and left for college, the Boy Wonder popped in now and then to help Batman solve a case. What makes a good sidekick different from a puppy dog is what I call the three Bs: brains, bravery, and brawn. These characteristics help him solve clues, save citizens, and defeat bad guys. These characteristics make a useful crime fighter.

To measure the three Bs, I'll take lessons from baseball and game shows. These are just two of the many places probability appears every day. Probability is the mathematical foundation of statistics, the toolkit from which all data analyses are built. The term is commonly used in two different ways, and it's important to know the difference. *Probability*, the mathematical language of uncertainty, describes what are called *random experiments*, bets, campaigns, trials, games, brawls, and anything other situation where the outcome isn't known beforehand. *A probability* is a fraction, a value between zero and one that measures the likelihood a given outcome will occur. A probability of zero means the outcome is virtually impossible. A probability of one means it will almost certainly happen. A probability of one-half means the outcome is just as likely to occur as not.

There are two types of probabilities: theoretical and empirical. Theoretical probabilities are constructed from logic and mathematical reasoning alone. These types of probabilities are often applied to gambling and game shows. For example, imagine a game show where the host shows you three closed doors. One of the doors hides a big pile of cash. The other two are empty. You pick a door, and the host opens it. What's the probability you choose the winning door? Three doors. One guess. Your chances of winning are one out of three, or 1/3. No data required.

Empirical probabilities are calculated from data. Empirical probabilities appear everywhere, in political polls, drug studies, weather predictions, financial reports, and sports. In baseball, for example, a player's batting average is an empirical probability. The batting average estimates the probability a player will get a base hit the next time he goes up to bat. This value is calculated as the number of base hits the player has made up until this point in the season (or career), divided by his total number of times at bat. A batting average of 0.300 means that the player has a base hit, on average, three out of every ten times at bat. That's a probability of 0.3 his next round at bat will result in a base hit.

So, what's Robin's batting average with respect to the three Bs? What's the probability he'll solve the next clue, rescue Batman from a jam, or deliver a decisive blow to the bad guy? There's no logic or mathematical reasoning that can answer this question (at least none that I know of), and so the only way to find out is to use empirical probabilities. This means I need to read Batman and Robin comics, lots of them, and keep track of these statistics. And that's just what I did. I chose the original comics from the Golden Age, 1940 to 1969, before Robin graduated high school and went off to college, while he was still a constant fixture in Batman's life.

Batman and Robin appeared together in hundreds of stories during these years, and I don't have the time or money to collect and read every single one of them. So, I did what every respectable statistician does when faced with the same problem. I sampled. Sampling is the process of choosing a sample, a subset of the population for data analysis. Random sampling, choosing members of the dataset at random, is popular because it eliminates the possibility of biases caused by handpicking your data. For example, I like the Riddler, an evil mastermind who often leaves riddles for clues. If I were to hand-pick Batman comics to read, I'd undoubtedly end up with quite a few featuring this villian. However, this strategy would probably leave me with an overinflated view of Robin's brains. For, while Batman seems to do most of the clue solving, Robin appears to have a knack for unraveling riddles. This being the case, my analysis would tend to overestimate Robin's clue-solving capacity. Random sampling prevents this from happening.

Random sampling would be the best way to select which comics to read. Unfortunately, getting my hands on a large number of randomly selected, sixty-year-old comic books is virtually impossible. Instead, I used the next best thing, three recently published collections designed to illustrate the breadth and variety of the Batman in the Golden Age. These collections, *Batman in the Forties*, *Batman in the Fifties*, and *Batman in the Sixties* (DC Comics 2004, 2002, and 1999) include over forty capers featuring the Dynamic Duo during this era. There may be an unforeseen bias in my data and it isn't perfect, but the real world never is.

Thanks, DC Comics, for doing my sampling for me.

PROBABILITY: THE SUPERHERO'S TOOLKIT

Let's return, for the moment, to theoretical probabilities, those probabilities that don't require data to calculate. To construct such a probability, you need three things: a random experiment, a sample space, and an event. The random experiment, or *trial*, is the situation whose outcome is uncertain, the one you're watching. A coin toss is a random experiment, because you don't know beforehand whether it will turn up heads or tails. The *sample space* is the list of all possible separate and distinct outcomes in your random experiment. The sample space in a coin toss contains the two outcomes heads and tails. The outcome you're interested in calculating a probability for is the *event*. On a coin toss, that might be the case where the coin lands on heads.

To calculate the probability of an event, simply count up the number of ways it can occur, and divide by the total number of outcomes in your sample space. In the case of a coin toss, there's one way a heads can occur and two possible outcomes, so the probability of a heads is 1/2. Calculating a probability in this way, by dividing the number of outcomes in your event by the total number of outcomes in the sample space, is a *classical probability*. In particular, for a random experiment with N possible outcomes, and an event E with M distinct outcomes, then the probability of E is

$$P\{E\} = \frac{M}{N}.$$

Calculating classical probabilities is straightforward, but not always easy. For example, imagine a game show with two rounds. In the first round, you answer trivia questions in order to win chances at the Wheel of Prizes. There are five prizes on the wheel, one of which is a car. What's the probability you'll be driving away in that new car? Consider

Bracelet, bracelet	Television, sofa	$500, $500
Bracelet, television	Car, bracelet	$500, sofa
Bracelet, car	Car, television	Sofa, bracelet
Bracelet, $500	Car, car	Sofa, television
Bracelet, sofa	Car, $500	Sofa, car
Television, bracelet	Car, sofa	Sofa, $500
Television, television	$500, bracelet	Sofa, sofa
Television, car	$500, television	
Television, $500	$500, car	

Figure 4.1. All possible winnings: two spins of the wheel of prizes.

two scenarios. First, suppose your trivia skills are merely average and you win a single spin of the Wheel of Prizes. In this case, your random experiment is a single spin of the wheel. Your sample space consists of all five prizes on the wheel. Your event is the wheel landing on the one spot with the picture of a car. The probability you'll win the car is one chance out of five spots, or 1/5. Pretty simple.

Now suppose you're a trivia wizard and you win two chances to spin the wheel. In this case, your random experiment is two spins of the Wheel of Prizes. Your sample space consists of all possible combinations of outcomes on two spins of the wheel, for example, a diamond bracelet on the first spin and a television on the second, or a television on the first spin and a car on the second. These outcomes are listed in Figure 4.1. (Note that a diamond bracelet on the first spin and a television on the second is a different outcome than a television on the first spin and a diamond bracelet on the second. The order of the prizes matters here.) If you count all the ways you can win the car, you'll find nine of them. (Again, a car on the first spin and a bracelet on the second is a different way to win than a bracelet on the first and a car on the second). So, the probability you'll win the car is 9/25.

Listing outcomes and counting them up may be doable for this example, but it's tedious. It rapidly becomes impossible as the size of your sample space grows. For example, suppose Batman receives a gift from his arch-nemesis, the Joker. It contains two harmless-looking die, six sided with dots printed on them, just like the ones from a typical board game. Of course, these are no ordinary die. Upon contact with a tabletop, they explode, releasing a hundred tiny capsules into the air. Each of these capsules holds either poisonous gas or a harmless bit of confetti. The sample space in this random experiment consists of every possible combination of a hundred poisonous and confetti particles. That's roughly 10^{30} combinations, too many for even the most diligent crime fighter to count!

Fortunately, we have what are called *counting rules*. Counting rules provide formulas for adding outcomes of common events and sample spaces. The formulas can get a little complicated, so I've banished those to Figure 4.2. It's possible you may never need any counting rules in your data analysis life, but a basic understanding of how different outcomes are counted can make probability calculations much easier, and so I feel compelled to spend a little time discussing them.

Take two successive spins of the Wheel of Prizes, for example. These two spins are what we call independent trials. *Independent trials* are repeated random experiments, where the outcome probabilities for any one trial are not affected by what happened before it. In other words, each trial is *independent* of all previous trials. More on independence shortly, but with respect to the Wheel of Prizes, this means that on the second spin of the wheel, your probability of landing on the car (or any other prize) is 1/5, regardless of what happened on the first spin. This is a common type of experiment, and there's a counting rule for adding up the total number of possible outcomes in such a case. So, rather than listing out all combinations of prizes, you can refer to Table 4.2 and use the formula. For N independent trials, where trial 1 has M_1 outcomes, trial 2 has M_2 outcomes, and so on, the total number of outcomes is $M_1 \times M_2 \times \cdots \times M_N$. In other words, the total number of outcomes across all trials is the product of the number of outcomes in the individual trials. For two spins of the Wheel of Prizes, that's 5×5, or 25.

The counting rules in Table 4.2 apply to other types of experiments as well. Suppose, rather than directing you to the Wheel of Prizes, the host holds out five envelopes, each containing a slip of paper with the name of a prize hidden inside it. You are asked to select two envelopes, one at a time. After your first draw, the host reveals your prize: $500. On the second draw, the probability of choosing $500 drops to zero—that envelope is gone—but the probability of the remaining prizes increases to 1/4. In other words, the two draws are not independent trials, because the probabilities on the second draw are impacted by the outcome of the first draw.

Rule	Formula	Uses	Examples
Independent Trials	N trials, where trial 1 has M_1 outcomes, trial 2 has M_2 outcomes, and so on, the total number of outcomes is $M_1 \times M_2 \times M_N$	Independent trials, sampling with replacement	Successive coin tosses, dice rolls, roulette wheel spins, pulls of a slot machine
Permutation Rule	*When order is important*, the number of ways you can draw N items from a collection of M items: $M \times (M\text{-}1) \times (M\text{-}N\text{+}1)$	Sampling without replacement	Drawing numbers from a hat, dealing cards from a single deck, selecting numbers for Bingo
Combinations Rule	*When order is not important*, the number of ways you can select N items from a collection of M items is M choose N, $$\binom{M}{N} = \frac{M!}{(M-N)!\,N!}$$ where $M!$ is $M \times (M\text{-}1) \times 1$.	Sampling without replacement	

Figure 4.2.   Common counting rules for classical probabilities.

This is an example of what we call *sampling without replacement*. Sampling without replacement occurs when you successively draw items from a collection without replacing the ones you drew previously. Sampling without replacement is the opposite of *sampling with replacement*, where items are replaced between draws. (And if you're following me, you might be able to see that sampling with replacement is just another name for independent trials.)

There are two commonly used rules for counting outcomes when sampling without replacement. The first rule, the *permutation rule*, counts all the possible ways your successive items can be drawn. This rule should be used when the order in which items are drawn matters. Referring to Figure 4.2, the total number of ways you can draw N items from a collection of M items without replacement is $M \times (M - 1) \times \ldots \times (M - N + 1)$. For two draws from the prize envelopes, where the first draw has five possible prizes to win and the second draw has only four, the number of total outcomes is $5 \times 4 = 20$.

The second rule, the *combination rule*, counts all the ways some combination of items can be drawn from the collection. This rule should be used when order does not matter. For example, the number of ways you can choose two prize envelopes out of five, without regard to which prize is first and which second, is *five choose two*, or

$$\binom{5}{2} = \frac{5!}{(5-2)!\,2!} = \frac{5 \times 4 \times 3 \times 2 \times 1}{(3 \times 2 \times 1) \times (2 \times 1)} = 10.$$

In other words, there are ten possible prize combinations you can win when you draw two envelopes from the game show host's hands.

Set Arithmetic 101: The Union, the Intersection, and the Complement

The classical probability rule is the foundation of statistics. It allows you to calculate the probability of almost any event you'd like. But by itself, this rule is like the Batmobile without its rocket blasters. It doesn't take you very far, or very fast. Most of the interesting probabilities involve combining more than one event. For example, suppose the evil mastermind King Tut plants a statue of the Egyptian god Anubis in the middle of Gotham Central Park. You and I both know the statue's been rigged to release deadly gas on curious onlookers, and so this event isn't terribly interesting. However, combine this with the event Batman and Robin show up to foil his evil scheme and now you've got something. Will the dynamic duo arrive before King Tut poisons his first innocent bystander? Calculating this probability requires us to combine events, something the classical probability rule just can't do alone.

Events are just sets, collections of outcomes in a sample space, and to combine events, you need some basic set arithmetic. To illustrate this arithmetic, it's useful to use a Venn diagram. A *Venn diagram* is just a box with circles inside it. The box represents the entire sample space of

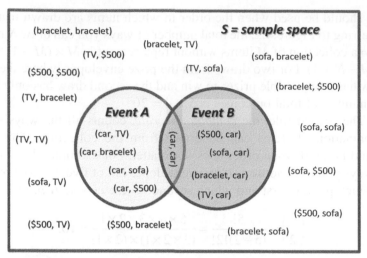

Figure 4.3. Venn diagram for two spins of the Wheel of Prizes.

a random experiment. The circles represent events that can occur. Returning to the game show scenario on more time, Figure 4.3 shows a Venn diagram for two spins of the five-slotted Wheel of Prizes. There are 25 outcomes in this sample space, as indicated in the figure. The circle A, shaded in light gray, represents the event you win a car on the first spin of the wheel. The circle B, shaded darker, represents the event you win a car on the second spin of the wheel. The probability of event A is simply the number of outcomes inside the region A divided by the total number of outcomes in S. There are five events in circle A. So, the probability of winning the car on the first spin, P{A}, is 5/25. Similarly, the probability of winning the car on the second spin, P{B}, is 5/25.

There are two ways to combine events A and B: the union and the intersection. The *union*, denoted by the mathematical symbol $\cup$, is the collection of all events in either A or B or both. On a Venn diagram, $A \cup B$ is represented by the area enclosed by both circles. In Figure 4.3, $A \cup B$ is the event that you win a car on either the first or second spin. In other words, it's the event you win at least one car. The probability of this union, P{$A \cup B$}, is simply the number of outcomes found inside this area divided by the total number of outcomes in S. Using Figure 4.3, the probability you win a car on two spins of the Wheel of Prizes is P{$A \cup B$} = 9/25.

The *intersection*, denoted ∩, is the collection of events residing in both A and B at the same time. This is the region between two circles on a Venn diagram. In Figure 4.3, $A \cap B$ is the event you win a car on the first spin and on the second spin. In other words, it's the event you win two cars. The probability of this event is the number of outcomes in both A and B divided by the total number of outcomes in the sample space. There's only one way to win the car on both spins of the wheel, so in this example, P$\{A \cap B\}$ = 1/25.

The union and the intersection are common in statistics, and these operations give us a way to combine two or more events. The complement is another important concept. For a set A, the complement of A, denoted A^C, includes all of the outcomes not in A. On a Venn diagram, this is represented by the entire region outside the circle A. In Figure 4.3, the event A^C is the event that you *do not* win the car on the first spin of the wheel. The probability of A^C can be calculated by counting all the outcomes outside A and dividing by the total number of outcomes in the sample space, however, if you already know P$\{A\}$, there's an easier way to find P$\{A^C\}$. Because the probability of the entire sample space is one, the probability of A^C is just one minus the probability of A. In other words, P$\{A^C\}$ = 1 − P$\{A\}$. So, the probability you do not win the car on the first spin of the wheel is P$\{A^C\}$ = 1 − 5/25 = 20/25, or 4/5.

INDEPENDENT AND MUTUALLY EXCLUSIVE EVENTS

There are many rules for calculating probabilities of combined events. The most common ones focus on two types: mutually exclusive events and independent events. Mutually exclusive and independent events are so important to statistics and so often confused it's worth spending a little time getting to know them.

Mutually exclusive events are events that cannot both happen at once. Think of the event Batman rushes into Police Commissioner Gordon's office, responding to the Bat-Signal. Now think of the event millionaire Bruce Wayne rushes into the same office to tell the commissioner his precious rare book collection has been stolen. Because Bruce Wayne is Batman's secret identity. In other words, they're the same person. So if you discount the possibility one of them is an imposter, both events cannot happen at the same time. These two events are mutually exclusive.

> Two events, A and B, are mutually exclusive if their intersection is empty. Mathematically speaking this means, $A \cap B = \varnothing$.

Two events are *independent* if the occurrence of one has no impact on the other, probabilistically speaking. For example, suppose two of Gotham City's worst villains are making plans. Catwoman hopes to steal the famous Cataran diamond from Spiffany's Jewelry store. Egghead wants a golden fossilized dinosaur egg from the Gotham City Museum of Natural History. Neither villain attended last month's Enemies of Batman meeting, and both forgot to send out a memo informing the city's other arch-criminals of their plans. In other words, they're operating independently of one another. Since neither knows the other's plans, the probability Egghead will hatch his plot on Tuesday is unaffected by the day Catwoman chooses to carry out her scheme. The probability Catwoman hits the jewelry store on Tuesday is the same whether or not Egghead hits the museum on the same day. Statistically speaking, the two events are independent.

Two events are independent if the probability of their intersection (the set of outcomes in both events) is the probability of the first event times the probability of the second. Here's the formal definition:

> Two events, A and B, are independent if $P\{A \cap B\} = P\{A\}P\{B\}$.

It's easy to confuse mutually exclusive and independent events. Both concepts suggest events that are separate and distinct from one another in some way. But they are not the same thing. Let me repeat, mutually exclusive events and independent events are *not* the same thing. For example, the event Egghead hatches his plot on Tuesday is independent of the event Catwoman carries out her plan on the same day. The probability of one is unaffected by whether or not the other occurs. But they're not mutually exclusive. Both events can happen at the same time.

On the other hand, Bruce Wayne can rush into the commisioner's office. Batman can rush into the commissioner's office. But both cannot happen at the same time. These two events are mutually exclusive. But they're not independent. If Bruce Wayne rushes into the commissioner's office, the probability Batman will also appear drops to zero. So the probability Batman will appear is strongly impacted by the appearance of Bruce Wayne.

In general, the following can be said about mutually exclusive and independent events:

> *If two events have nonzero probability, then they cannot be both independent and mutually exclusive at the same time.*

Probability Rules

When combining events, statisticians manipulate, redefine, and otherwise work very hard to make them either mutually exclusive or independent. The reason is simple. There are lots of rules for combining these special types of events. As a general rule, if you need to calculate the probability of a union, the accumulation of all outcomes in two events, it's nice to have mutually exclusive events. If you need to calculate the probability of an intersection, those outcomes residing in two events at once, it's nice to have independent events. Here's why.

> *Addition rule:* For two mutually exclusive events, A and B, the probability of their union is the sum of the individual probabilities. Mathematically speaking, this means $P\{A \cup B\} = P\{A\} + P\{B\}$.

> *Multiplication rule:* For two independent events, A and B, the probability of their intersection is the product of their individual probabilities. In other words, $P\{A \cap B\} = P\{A\}P\{B\}$.

Suppose you want to calculate the probability that either Batman or Bruce Wayne will be the next person to enter Commisioner Gordon's office. This is the probability of the union of two events: (1) Batman will be the next one to enter Commisioner Gordon's office, and (2) Bruce Wayne will be the next one to enter Commisioner Gordon's office. Since they can't both happen at the same time, these two events are mutually exclusive. So, if B is the event Batman appears and W is the event Bruce Wayne appears, then $P\{B \cup W\} = P\{B\} + P\{W\}$.

Now suppose you'd like to calculate the probability Gotham City will see a crime from both Catwoman and Egghead next Tuesday. This is the probability of an intersection of two events: (1) Catwoman carries out her latest plan on Tuesday, and (2) Egghead hatches his plot on

Tuesday. If these two events are independent, the probability both happen on the same day is just the individual probabilities multiplied together. If C = the event Catwoman carries out her plans on Tuesday and E = the event Egghead hatches his plot on Tuesday, then $P\{C \cap E\} = P\{C\}P\{E\}$.

Conditional Probabilities

I have a rule of thumb I apply to all my statistical analyses. If you have information, use it. For example, say you're a huge comic book fan and you'd like to calculate the probability Batman will walk through your front door in the next ten minutes. Knowing Batman's true identity, your estimate would be dramatically different if Bruce Wayne were standing next to you than it would be if you were alone in the room. Statisticians use *conditional probabilities* to incorporate knowledge such as this into their probability calculations.

A conditional probability, $P\{A|B\}$, is the probability event A will occur given you know event B has occurred. For two events A and B, the probability of A *given* B is the probability both A and B occur (their intersection), divided by the probability B occurs. Mathematically,

$$P\{A|B\} = \frac{P\{A \cap B\}}{P\{B\}}.$$

In general, a conditional probability can be any number between zero and one, depending on how much the two events impact one another. At one extreme, if A and B are mutually exclusive, they cannot both occur at the same time and so the conditional probability of A *given* B is zero. For example, the probability Batman will walk through your front door is zero when you have Bruce Wayne sitting on your living room couch. At the other extreme, if A and B are independent, then when one occurs, it has no impact on the probability of the other occurring and so the probability of A *given* B is just the probability of A. This means the probability Catwoman will strike today is the same as it was before you learned that Egghead just robbed the museum of its golden eggs. Between these two extremes lie the non-mutually exclusive, non-independent events. These events influence one another in more subtle ways.

For example, suppose Batman does walk through your front door. He can appear alone or with his sidekick, and his sidekick can be in superhero form, as Robin, or in civilian form, as Dick Grayson. None of these scenarios is impossible. So, letting B represent the event that Batman appears and R represent the event that Robin (and not Dick Grayson) appears with him, these two events, B and R, are not mutually exclusive. Nor are they independent. Batman prefers to travel with his sidekick in superhero, not civilian form. This means the probability Robin will appear increases the moment you see Batman enter the room. In other words, the conditional probability that Robin appears given Batman has just appeared is higher than it would be if you hadn't taken your knowledge of Batman into account.

Let's do a calculation using completely made up probabilities. Suppose the probability Batman will stop by for a visit today is 0.25, and the probability Robin will visit is 0.05. Suppose also the probability both Batman and Robin will visit together is 0.1. Assuming Batman visits, the probability he'll bring Robin with him is the conditional probability $P\{R|B\}$.

$$P\{R|B\} = \frac{P\{R \text{ and } B\}}{P\{B\}} = \frac{0.1}{0.25} = 0.4.$$

So, given Batman stops by, there's a 40% chance he'll bring Robin along. This is much higher than the probability would be if the two events were independent. (In which case, $P\{R|B\}$ would be the same as $P\{R\}$, or 0.05.)

Empirical Probability: For Those Grounded in Data

Theoretical probabilities, like those presented in the previous section, are the foundation of statistics. Not only do they give us a way to calculate likelihoods for simple random experiments, they also lay out the rules for more complicated ones. In the real world, people are typically more interested in empirical probabilities than theoretical probabilities. Empirical probabilities are estimates, values that (1) are calculated from a sample of data and (2) are meant to approximate the corresponding probability from the population under study. They appear everywhere, either by themselves or masked as percentages and odds. Relative frequencies and the frequency distribution from the previous chapter are empirical probabilities. Political polls report the percentage of respondents who favor a particular candidate, this percentage reflecting the empirical probability a random voter would choose that candidate. Drug commercials often report the risk of certain drugs, this risk being a ratio of empirical probabilities. Empirical probabilities also appear in sports. Baseball, in particular, is full of them.

Empirical probabilities are calculated using the classical probability rule and the experimental design process. First, you decide on a random experiment and then define the sample space and the event you're interested in. Then you watch many independent trials of this experiment, keeping track of the number of times your event occurs. Divide this number by the total number of trials in your experiment and you're done. In baseball, a player's batting average is calculated in this way. The random experiment is a player's time at bat. The sample space consists of two outcomes: a base hit or something else (an out, a walk, etc.). The event of interest is a base hit. The batting average is the player's total number of base hits divided by the number of times at bat. This batting average can be calculated for a single season or for the player's entire career. Consider a player who's been up to bat twenty times this season. If he's made a base hit five of those times, his season batting average is 5/20, or 0.250.

More complex empirical probabilities, those involving more than one event, can be combined in the same way theoretical probabilities are combined. For example, a baseball player's batting average against a given pitcher is sometimes reported by announcers. This is a conditional probability, $P\{H|P\}$, where H = the event the player will get a base hit and P = the pitcher in question. In other words, it's the probability a given player will get a base hit given P is pitching. To calculate this, you can apply the definition of conditional probability to your calculations. Specifically, you count up the number of times a player has a base hit when pitcher P is pitching and divide that by the number of times the player has gone up against pitcher P. If a given player has gone up against the pitcher ten times this season, and three of those times he's gotten a base hit, his current season batting average against that pitcher is 3/10, or 0.300.

SIDEKICKS ARE PEOPLE, TOO

I wish I had a Batcave. It's an underground complex where Batman and Robin do much of their crime solving. It houses the Batmobile and the Batplane. There's an underground laboratory, a workshop, and even a trophy room where the superhero keeps some pretty strange criminal devices he's collected over the years. My favorite room houses the computerized crime file. I can only imagine what this wonder of technology does, but in my mind, it keeps track of everything going on in Gotham City—news, weather, social events—and distills all that information down to what's important for an investigation. It also performs voice-activated data analysis and makes a great cup of espresso.

Since I don't have my own computerized crime file, I had to rely on less sophisticated methods to calculate Robin's batting average. First, I read through every comic in *Batman in the Forties*, *Batman in the Fifties*, and *Batman in the Sixties*, counting the number of times Batman, Robin, or both together exhibited one of the three Bs. In other words, I kept track of how many times each of them either solved a clue, rescued themselves from a criminal's trap, or punched a bad guy. I also counted the total number of clues, traps, and punches in all of the comics, and threw these numbers at Microsoft Excel.

There are two common ways to display empirical probabilities. When there are many categories or outcomes in a sample space, the relative frequency distribution is often displayed with a bar chart or a histogram. You can find lots of these types of charts in Chapter 3. When there are relatively few categories or outcomes to plot, a pie chart works nicely. A pie chart shows the different outcomes as slices of a pie, where the size of each slice represents the relative proportion of a given outcome. Figure 4.4 displays the pie charts for each of the three Bs— brains, bravery, and brawn—as measured by (a) clues solved, (b) rescues, and (c) punches delivered.

In each chart, Batman is shaded light, and Robin is shaded dark. Clues or rescues needing both Batman and Robin are shaded in medium gray between the two extremes. In the pie charts, the relative proportions are reported as percentages, but since a percentage is simply a fraction multiplied by one hundred, they can easily be turned into a probability simply by moving decimal point two places to the left.

In these three pie charts, the shading alone makes it clear. Batman dominates in all areas. He alone solves the vast majority of the clues and delivers most of the punches. He alone performs almost half of the rescues. He alone is the superhero. On the other hand, Robin's contribution cannot be ignored. He delivers 40% of the blows, and so his brawn batting average is P{Robin will deliver a punch} = 0.4. Furthermore,

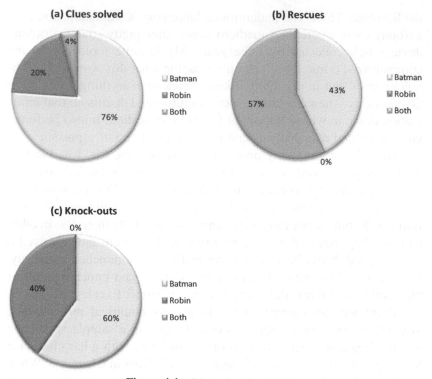

Figure 4.4.   Measuring the three Bs.

he may not perform any of the rescues by himself, but Batman needs him 57% of the time to get out of a jam.

Robin's biggest weakness appears to be in the brains department. He solves a mere 4% of clues by himself, and helps Batman an additional 20% of the time. The other 76% of the time, it's up to Batman to do the heavy brain work. However, as I mentioned at the beginning of this chapter, the Boy Wonder does seem to have a knack for solving the Riddler's clues. There was only one Riddler story included in my initial dataset, but in that story, Robin unravels three out of six riddles.

Could solving riddles be one of Robin's many talents? To find out, I tried to get my hands on a bunch of Batman and Robin capers involving the Riddler. Easier said than done. As it turns out, the Riddler appeared in only ten Batman and Robin comics between 1940 and 1970. Fortunately, with the help of my local comic book store, I was

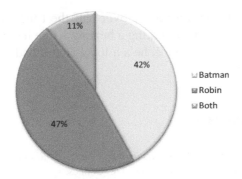

Figure 4.5. Batman and Robin vs. the Riddler: clues solved.

able to get a total of five of them. One story was from my original collection, and four more were available in *Batman: Featuring Two-Face and the Riddler* (DC Comics, 1995), and *Showcase presents Batman*, vol. 3 (DC Comics, 2008).

There were a total of nineteen riddles in these five capers. Most of these riddles had an obvious answer and a more subtle meaning. Generally, either Batman, Robin, or both would answer the riddle, and then both would unravel what it meant to their investigation. Figure 4.5 displays a pie chart of Batman and Robin's record in solving these riddles. Because only capers involving the Riddler are included, this pie chart represents the conditional probabilities that Batman, Robin, or both will solve a riddle. According to the figure, Robin alone solves 47% of them, giving him a riddle-solving batting average of P{*Robin will solve a clue | the Riddler is the villain*} = 0.47. A batting average like this qualifies him for the All-Stars. It also shows that the event *Robin will solve a clue* is not independent of the event *the Riddler is the villain*. If they were independent, then the Boy Wonder's batting average against this villain would be the same as his overall batting average, 0.040. Instead, it's much higher.

NEVER LEAVE HOME WITHOUT ONE

On the whole, Robin may not be the brains of the dynamic duo, but he makes up for it in other ways. He's a real asset when the Riddler's in town. He follows Batman everywhere, sticking by his side in a pinch, jumping in to help with a rescue over half the time. And he's always ready to fight for his friend, taking a beating while he dishes out almost half of the punches. All in all, he's a loyal friend and worthy sidekick. Just like probability.

BIBLIOGRAPHY

DC Comics. 1995. *Batman: Featuring Two-Face and the Riddler*. DC Comics.
DC Comics. 1999. *Batman in the Sixties*. DC Comics.
DC Comics. 2002. *Batman in the Fifties*. DC Comics.
DC Comics. 2004. *Batman in the Forties*. DC Comics.
DC Comics. 2008. *Showcase Presents: Batman*, vol. 3. DC Comics.
MLB. "Official Info." http://mlb.mlb.com/mlb/official_info/baseball_basics/ stats_101.jsp?c_id=mlb, accessed May 3, 2012.

CHAPTER 5

What It's Like to Be Normal: Probability Distributions and One Rule That Could Make You Wildly Popular

Remember that girl from high school? You know the one. She had two happily married parents, 1.4 siblings, decent grades, and a pleasant but not too perfect overbite. She was friendly and wholesome and fun to be around. I always envied this girl with her easy manners and effortless hair (my social skills were awkward and my own hair always looked more like hay than silky golden locks). Even so, I couldn't help but like her. Everybody liked her.

Just like high school, statisticians have their own popularity contests. But what warms our hearts isn't a pleasant smile. It's power. The power to characterize data. The power to confidently state what we know and, just as importantly, what we don't know. The power to answer questions in the midst of life's uncertainty. This kind of power requires more than a few descriptive statistics and some counting rules.

The Art of Data Analysis: How to Answer Almost Any Question Using Basic Statistics,
First Edition. Kristin H. Jarman.
© 2013 John Wiley & Sons, Inc. Published 2013 by John Wiley & Sons, Inc.

It requires a tool for calculating the probabilities of outcomes in a random experiment, a tool that can be used on any type of data, no matter how complex. This kind of power requires something called the probability distribution.

There are many types of probability distributions out there, and each one has its uses. Without a doubt, no probability distribution is more common or more popular than the normal distribution. This girl-next-door of all probability distributions is well-rounded and easy to work with, and many types of data naturally conform to its pleasing personality. As if that's not enough, there are two rules that can mold almost any set of data, no matter how awkward, into the normal distribution. Not only does this make the normal distribution wildly popular among statisticians, it also gives all of us nerds, geeks, outcasts, and oddballs out there a chance to answer one of our most secret, burning questions.

What's it like to be normal?

THE GRANDDADDY OF ALL DEMOGRAPHIC DATA SOURCES

If you live in the United States, you probably know about the Census Bureau. This government agency is responsible for keeping track of the demographics of the U.S. population: age, income, housing, family size, and so on. Every ten years, the Census Bureau takes on the monumental task of surveying of every man, woman, and child in the United States and compiling this information into a massive dataset capturing not a sample, but the entire population of the country. The government uses these data to predict trends in income, lifestyle, and population distribution so it can set policies and allocate resources accordingly. The Census Bureau makes these data publicly available to anyone who wants them. As of this writing, data from the most recent decennial survey, the 2010 Census, is gradually becoming available on their website.

Most people know about this massive data collection effort that takes place every ten years, but that's not the end of it. Big Brother is always watching, and the U.S. Census Bureau is constantly collecting information on the population. In fact, this agency conducts six ongoing surveys covering everything from emergency care to salaries to commuting time (U.S. Census Bureau 2012), and at any given time, at least one of these surveys is in progress. All of this information is available on their website, www.census.gov. You only need a little time and some determination to get it.

The American Community Survey is one of the ongoing surveys conducted by the Census Bureau. Each year, a small percentage of the population is asked to provide information about home and work life. This includes income, work benefits, disabilities, community involvement, and commuting time. It's the perfect dataset for an oddball like me, the perfect way for a statistician to learn all about normal. Without a dataset like this, I'd need to design my own survey. I'd need to decide what questions to ask and how to ask them in an objective way. I'd need to decide how many people to call and select those people at random from the entire U.S. population. But thanks to the U.S. Census Bureau, that's already been done for me. And it's been done by the experts.

If you're looking for a class project idea or if you're merely curious, I encourage you to navigate to the Census Bureau's website and do a little data shopping. I will warn you, however, there's a lot of information out there. If your computer tends to get queasy at the sight of large files, avoid downloading the detailed data and stick to summaries instead. For my assessment of normal, I downloaded every record from the 2010 American Community Survey, this relatively small annual sampling of the U.S. population. The result was over 98,000 records. I love data as much as the next statistician, but there can be too much of a good thing!

PROBABILITY DISTRIBUTIONS AND WHY WE NEED THEM

Based on descriptive statistics—sample means, medians, and modes—pulled from recent data summaries on the U.S. Census Bureau's website, the typical American is a 44-year-old woman named Smith, a wife and mother who drives to work alone, makes about $45,000 a year, communicates over the Internet virtually every day, talks to her neighbors a couple times a week, and never discusses politics.

If that's you, then congratulations, you're officially normal.

Of course, most of us don't fit this exact description. We're the others, the ones who take the train to work, the ones who hate our neighbors *and* the Internet, the ones who freely share our political opinions with anyone who'll listen. So what does this description mean to all of us? Does it mean we're outcasts doomed to spend our freakish lives on the fringes of society? Worse yet, does it mean we don't count in the big scheme of all things demographic? The answer to these questions can be found in something called the probability distribution.

A *probability distribution* is a mathematical function that attaches probabilities to all possible outcomes in a random experiment. You've already seen simple probability distributions such as the one for a coin toss, where

$$P\{heads\} = \frac{1}{2}, \quad and \quad P\{tails\} = \frac{1}{2}.$$

This distribution gives us the probability that the coin will land on either one of its two sides. All you need to do is look at the function value for the outcome you're interested in.

If all random experiments were as simple as a coin toss, we wouldn't really need probability distributions. We could rely on the classical method and counting rules to calculate any probability we wanted. However, the real world rarely gives us such simplicity. From calculating sales trends to estimating the reliability of pacemakers, most real data analyses require more sophisticated tools for characterizing random experiments. The probability distribution is just such a tool.

The foundation of any probability distribution is a random variable. Usually denoted by a capital letter such as X, a *random variable* is nothing more than a mathematical value describing the outcome of a random experiment. For example, think of a coin toss. In the last chapter, without the use of random variables, I simply called the outcomes {heads} and {tails}. Now, I can become slightly more mathematical and say "let X be the outcome of a coin toss, where 1 refers to heads and 0 refers to tails." In this case, $P\{X = 1\} = 1/2$.

By representing outcomes as random variables rather than wordy event descriptions, we can be more precise, more versatile, and more efficient. Suppose you toss a coin three times and you'd like to know the probability at least two of those tosses come up heads. Without the use of random variables, you'd have to do something like the following:

P{at least two of three tosses are heads}

= P{1st and 2nd toss heads, 3rd toss tails}

+ P{1st toss heads, 2nd toss tails, 3rd toss heads}

+ P{1st toss tails, 2nd and 3rd tosses heads} + ⋯.

and so on for all possible combinations resulting in two or three heads. This formulation is much simpler with a random variable. In particular, if we let X_n be the outcome of the nth coin toss, where $X_n = 0$ (tails) or 1 (heads), then we can simply say

P{at least two heads} = $P\{X_1 + X_2 + X_3 \geq 2\}$.

This second formulation is more mathematical and takes a little getting used to, but as you'll see throughout this chapter, it's much easier to work with when the random experiments become more complicated and more realistic.

There are three important requirements for the probability distribution. First, it should be defined for every possible value the random variable can take on. In other words, it should completely describe the sample space of a random experiment. Second, the probability distribution values should always be nonnegative. They're meant to measure probabilities, after all, and probabilities are never less than zero. Finally, when all the probability distribution values are summed together, they must add to one. For example, the distribution for a coin toss could be formally described as follows:

$$P(X = x) = \begin{cases} \dfrac{1}{2} & \text{for } x = 0 \\ \dfrac{1}{2} & \text{for } x = 1. \end{cases}$$

This function meets all three requirements for a probability distribution function. It's defined for both values X can take on, both of the probabilities are nonnegative, and when added together, they sum to one.

Random variables and probability distributions bring all the power of mathematics to the business of characterizing random experiments. Random variables can be added, subtracted, and multiplied together as needed. They can have an unlimited range or infinitely many possible values, whatever the situation requires. Probabilities can be calculated with a simple formula, without the need for binning or counting outcomes. In short, random variables and probability distributions give us the ability to describe almost any type of random experiment, no matter how complex.

There are many well known probability distributions, and each one has a specific purpose. All of them fall into two basic categories: discrete and continuous.

Discrete Probability Distributions

Discrete random variables are used for experiments where the outcomes are discrete, or countable. A coin toss, the number of hits a website receives in a 24-hour period, the number of text messages you'll respond to in the next hour, all of these values are discrete random variables, because you can count up all the possible values in the sample space. *Discrete probability distributions* describe discrete random variables.

One of the topics covered in the 2010 American Community Survey is civic involvement. Respondents were asked many questions about their social habits, including how often they eat dinner with their family, how frequently they communicate over the Internet, and whether or not they've ever boycotted a product. Figure 5.1 plots the frequency distribution of responses to this last question. The most popular answer, mode

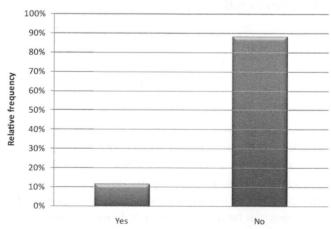

Figure 5.1.   Responses to the ACS 2010 question: Have you ever boycotted a product?

of this distribution, is "No." In other words, the vast majority of Americans have never boycotted a product.

Let X be a random variable, where $X = 0$ if a person has participated in a boycott and $X = 1$ if a person has not. This random variable, X, is a discrete random variable because you can count all the outcomes in the sample space, all two of them. According to Figure 5.1, 11.6% of people have boycotted a product. Rounding this value to 12%, the probability distribution function for X is the following:

$$P(X = x) = \begin{cases} 0.12 & \text{for } x = 0 \\ 0.88 & \text{for } x = 1. \end{cases}$$

This formula should look a lot like the probability distribution for a coin toss, with good reason. Both random variables follow the same probability distribution. It's the *Bernoulli distribution*, which describes a random variable with two possible outcomes and success probability (the probability $X = 1$) of p. The possible outcomes of the two random variables are the same. Only the probabilities are different.

All probability distributions have two parts: (1) a mathematical function, and (2) one or more parameters that go into the function. The function is a formula that determines possible outcomes and general tendencies of the random variable. This function is like a high school clique. It defines, very generally, how the random variable will behave. For example, jocks will play sports. Chess club kids will be smart. The Bernoulli distribution will have two possible outcomes. The *parameters* of a distribution dictate the specifics, things like the central location and variation. These parameters are like the individuals in a clique, each with his or her own unique personality. All jocks may play sports, but some might prefer basketball over football. All Bernoulli random variables have only two possible outcomes, but the probabilities of those outcomes may be very different.

Discrete probability distributions are generally plotted in the same way frequency distributions are plotted, with the possible values on the x-axis and the probability of each value on the y-axis. And like frequency distributions, probability distributions can be left-skewed, right-skewed, or symmetric. Figure 5.2 plots some of the most common discrete distributions. As you can see from the figure, their shapes vary. The binomial and Poisson distributions, for example, are almost symmetric and near bell-shaped. The discrete uniform distribution is also

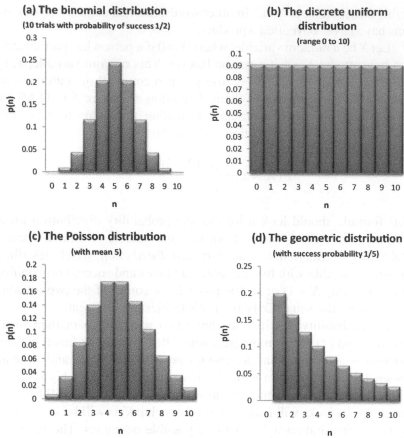

Figure 5.2.   Common discrete probability distribution functions.

symmetric, but its shape is rectangular. And the right-skewed geometric distribution, with its long tail, is neither symmetric nor bell-shaped. These different distribution shapes reflect different tendencies of a random variable, and each shape depends not only on the name of the probability distribution, but also its parameter values.

If you recall from the last chapter, independent trials are successive random experiments where the outcome of the current trial is not affected by the outcome of previous trials. One of the most common discrete probability distributions, the binomial distribution (Figure 5.2a), applies to this important situation. Specifically, the *binomial distribution* is a probability distribution describing the number of successes in a fixed set of independent trials. The parameters for the binomial distribution are the number of trials (N), and the probability of success on any one trial (p). For example, the number of heads you get on three coin tosses is a binomial random variable with parameter values $N = 3$ and $p = 1/2$. The mathematical formula for the binomial distribution is left to more rigorous textbooks, such as those listed in the reference section. For practical purposes, many spreadsheet software packages have functions for calculating probabilities of the binomial distribution. In Microsoft Excel, this function is called BINOMDIST.

The binomial distribution applies to any situation where you'd like to count specific occurrences: the number of "yes" answers on a survey, the number of defective parts coming off an assembly line, or the number of wins your favorite football team has in a given season. The binomial distribution can also be used to count responses in the 2010 American Community Survey. For example, one of the questions on this survey asks respondents how often they talk to their neighbors. There are five possible answers to this question, ranging from "Basically every day" to "Not at all." Figure 5.3 plots the frequency distribution of the responses.

The most common answer, the mode, is "A few times a week." So, if we were only looking at descriptive statistics, we'd say the typical American talks to his or her neighbors a few times a week. However,

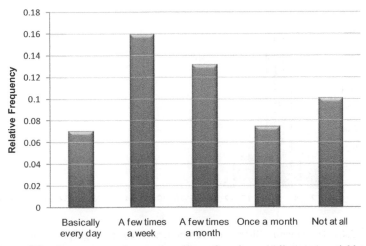

Figure 5.3.   Responses to the question: How often do you talk to your neighbors?

only 16% of people fit this description of what I'll call *moderately chatty*. The other 84% of us don't. We talk to our neighbors more or less frequently than a few times a week, and there's a lot more of us than of those perfectly typical people. In other words, the mode may be the most popular category, but it doesn't necessarily include most of the people.

If most individuals don't fit the typical profile of *moderately chatty*, what does this mean for an entire neighborhood? Suppose you have forty neighbors total. Because the probability any one of them will be moderately chatty is 0.16, the total number of *moderately chatty* neighbors you have, call it X, is a binomial random variable with number of neighbors $N = 40$ and success probability $p = 0.16$. Individual probabilities for different values of X can be calculated using the BINOMDIST function in Excel. The entire distribution function, made up of all of the individual probabilities, is plotted in Figure 5.4. Note the shape and placement of the distribution. The bulk of it sits on the low end of the range, far below forty and centered around six, but despite being a little off center, it has a nice, symmetric, almost bell-shaped distribution (more on this to come).

If *moderately chatty* is the typical neighborly behavior, then you'd expect many of your neighbors to fit this description, right? Let's find the probability more than a quarter of your neighbors, or ten out of

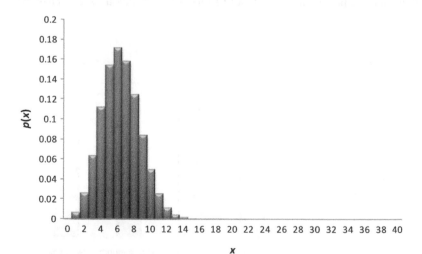

Figure 5.4. The binomial distribution for moderately chatty neighbors (40 neighbors with success probability 0.16).

forty, will be moderately chatty, in other words, $P\{X > 10\}$. This value can be calculated from the cumulative distribution function. The *cumulative distribution function*, or CDF, is simply the probability $P\{X \leq x\}$ for any value of x, and it's calculated by adding the probabilities for all values up to and including x. This value can be found the hard way by adding up the probability distribution function for all values below x, for example, $P\{X \leq 3\} = P\{X = 0\} + P\{X = 1\} + P\{X = 2\} + P\{X = 3\}$. However, this type of calculation isn't often necessary because most spreadsheet programs have built-in functions for calculating these cumulative probabilities. In Excel, the BINOMDIST function with the cumulative parameter set to TRUE will do this for you, and this function gives $P\{X \leq 10\} = 0.95$.

This value, $P\{X \leq 10\}$, isn't quite what we were looking for. We're interested in the probability *more than* ten neighbors will be moderately chatty. This value, $P\{X > 10\}$, can be calculated with a helpful little probability rule, the *complement rule*:

> For any value of x, $P\{X > x\}$ can be calculated from the cumulative distribution function using this formula:
>
> $$P\{X > x\} = 1 - P\{X \leq x\}.$$

As you'll see in upcoming chapters, the value $P\{X > x\}$ is used in many different types of statistical analyses, and so it's important to understand the complement rule. Remember, the probabilities for all possible outcomes of a random variable must add to one. The outcomes where $X \leq x$ and $X > x$ include all possible values of a random variable, and so their probabilities must also add to one. This relationship is shown in Figure 5.5.

The complement rule gives us an easy way to calculate the probability ten of your forty neighbors will be moderately chatty. Since $P\{X \leq 10\} = 0.95$, $P\{X > 10\} = 1 - 0.95 = 0.05$. In other words, in a neighborhood of forty people, there's only a 5% chance that more than ten of them will fit the typical, moderately chatty profile.

If more than ten is unlikely, then how many moderately chatty neighbors would you expect to have? In Chapter 3, I would have answered this question with descriptive statistics, by calculating the sample mean and standard deviation of a bunch of data collected from many different randomly selected neighborhoods. But now I can simply

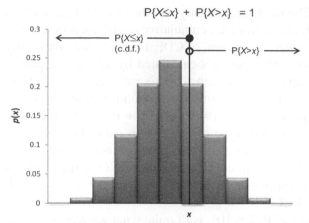

Figure 5.5. The complement rule.

use the probability distribution in Figure 5.4. Like datasets, random variables have a central location and variation associated with them, and these properties are used often. Denoted by the Greek letter μ, the *mean* describes a random variable's central location, in other words, the arithmetic center of the values a random variable can take on. Denoted with the symbol σ^2, the *variance* is the squared deviation of values around the mean, in other words, the standard deviation squared.

Conceptually, the mean and variance of a random variable are just like the corresponding values calculated from a sample of data. However, unlike descriptive statistics, these quantities are theoretical values, calculated directly from the probability distribution function. No data are required. The mathematical details on how to calculate the mean and variance of a random variable are left to texts more rigorous than this one. In Figure 5.6, I've listed formulas for the mean and variance of some common probability distributions introduced in this and the next section. These formulas are always the same for a distribution, but their values depend on the parameters chosen. In other words, two random variables can belong to the same group, such as the binomial distribution, and still have very different personality traits, like a mean of ten, a hundred, or even a thousand.

Back to the social habits of your neighbors. In a neighborhood of forty, how many moderately chatty neighbors would you expect to have? From Figure 5.6, the mean of a binomial random variable with N trials and success probability p is $\mu = Np$. So, the mean number of moderately chatty neighbors is $\mu = 40*0.16 = 6.4$. In other words, in a

Distribution name and type	Parameters	What it measures	The mean and variance
Bernoulli (discrete)	Success probability p	Any random experiment with two outcomes	$\mu = p$ $\sigma^2 = p(1 - p)$
Binomial (discrete)	Number of trials N Success probability p	The number of successes in N independent trials	$\mu = N_p$ $\sigma^2 = Np(1 - p)$
Poisson (discrete)	Success rate per unit time λ	The number of successes in a specified time interval	$\mu = \lambda$ $\sigma^2 = \lambda$
Exponential (continuous)	Expected time between events, λ	The time until a desired event occurs	$\mu = 1/\lambda$ $\sigma^2 = 1/\lambda^2$
Normal or Gaussian (continuous)	Mean μ Variance σ^2	Many different types of random experiments	$\mu = \mu$ $\sigma^2 = \sigma^2$

Figure 5.6. Common probability distributions.

neighborhood of 40 people, you'd only expect about six of them to fit the most typical chattiness profile. The rest would fit something else.

You might wonder how the mean of a random variable that only takes on integer values could be such an impossible, non-integer value as 6.4. This is common for discrete distributions and nothing to worry about. The mean of a random variable is an arithmetic value, just like the average. It describes the central location of some theoretical data cloud made up of all possible values for a random variable. For a discrete distribution, such a cloud contains holes, points on the real line where no possible outcomes lie, so it's quite easy to have an arithmetic center that falls into one of these holes. In other words, the mean doesn't need to hit one of the possible values, it only needs to fall inside the range of the probability distribution.

Continuous Probability Distributions

Continuous random variables are used for continuous observations. These are the variables whose possible values cannot be counted. The number of gallons of gas it takes to fill your car, the amount of medication needed to cure a disease, the concentration of pollutants in your local water source, all of these are random variables that can take any value in some range. And no matter what two values you choose, there will be more values between them. For example, between 15.4 and 15.5 gallons of gasoline is 15.45 gallons. Between 15.45 and 15.46 gallons is 15.455 gallons, and so on.

Continuous probability distributions describe continuous random variables. Like discrete distributions, they have two parts to them: a function describing general tendencies and parameters that specify the particulars. Like discrete distributions, they're used to calculate probabilities of specific values a random variable can take on. And like discrete distributions, each continuous distribution has a mean and variance associated with it. However, continuous and discrete probability distributions are not exactly the same thing for one very important reason.

For a discrete random variable X, the value of the probability distribution function at x is an actual probability, $p(x) = P\{X = x\}$. This isn't the case for a continuous random variable. Why? Remember the values of a continuous random variable cannot be counted. There are infinitely many of them, and if you choose two, no matter how close together, you can always find another in between them. Thinking in terms of the classical probability method, the probability of any one value is some number divided by the total number of outcomes. For a continuous random variable, the total number of outcomes is infinite and so the probability of any one outcome is zero.

On the other hand, a continuous random variable has to take on some value. How can the probability of every individual outcome be zero when one of those outcomes eventually occurs? Spending too much time thinking about this paradox can make you crazy, and unfortunately, the answer lies well beyond the scope of this book. For practical purposes, the only important thing to remember is that a probability density function doesn't give you probabilities of individual outcomes.

What does it give you? I like to think of it this way: a continuous probability distribution function gives you relative likelihoods of ranges of a continuous random variable, much in the way a histogram gives you relative likelihoods for bins of data. In particular, a random variable is more likely to fall in a range where the probability density function values are high, and least likely to fall in a range where they are low. More importantly, the probability density function is the foundation of the cumulative distribution function, $P\{X \leq x\}$. The cumulative distribution function for a continuous random variable *is* a legitimate probability, and it's used widely in statistics.

Perhaps because discrete and continuous distribution functions measure slightly different things, their names are different. For a continuous random variable, this function is called the *probability density function*. Requirements for a probability density function are as follows. It is always greater than or equal to zero. The sum total of the probability density function for all possible values of X is one (that's the integral of the function over all x, for those who have taken calculus). And the cumulative probability distribution $P\{X \leq x\}$ is the probability the random variable will be less than or equal to x. For most common probability distributions, the cumulative distribution function has either already been figured out for you, or can easily be calculated using common tables or formulas in programs like Excel.

Some common continuous probability density functions are plotted in Figure 5.7. As with discrete distributions, the shapes of these probability density functions reflect random variables with dramatically different tendencies. For example, the normal distribution is the perfect bell curve. Random variables with this distribution are more likely to fall in the center of the range, near the peak function value. The uniform distribution is symmetric but oddly square, and for mean three, it abruptly ends at a value of six. Random variables following the uniform distribution can occur anywhere in their range with equal probability. The exponential distribution is definitely not normal, with its right-skewed shape and stretched tail. Random variables having this distribution tend to fall on the low end of their range.

Far and away the most popular continuous distribution is the *normal distribution* (Figure 5.7a). Chances are, you've already run across the normal distribution somewhere. It's also called the Gaussian distribution or the bell curve. The normal probability density function is symmetric and nicely rounded. It has two simple parameters: the mean and the variance. This distribution is so important to basic statistics and so popular, I'm going to break my rule of no complicated formulas and show you what it looks like.

For a random variable X following a normal distribution with mean μ and variance σ^2, the *normal probability density function* is

$$p(x) = \frac{1}{\sqrt{2\pi}\sigma} e^{-\frac{(x-\mu)^2}{2\sigma^2}} \quad \text{for} \quad -\infty < x < \infty.$$

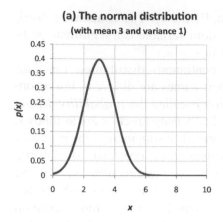

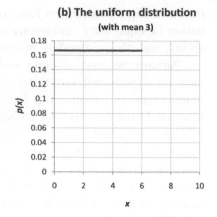

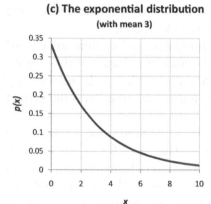

Figure 5.7. Common continuous probability density functions.

Put into words, a normal random variable can take on any real value, positive, negative, or zero. The density function is always largest at the mean, and it gradually decreases to zero on both sides. The width of the curve is determined by the variance parameter—the larger the variance, the shorter and wider the bell. Figure 5.8 illustrates how different mean and variance values change the central location and width of the bell curve.

Figure 5.8a plots a special case of the normal distribution, the *standard normal distribution*. In this case, the mean is zero and the

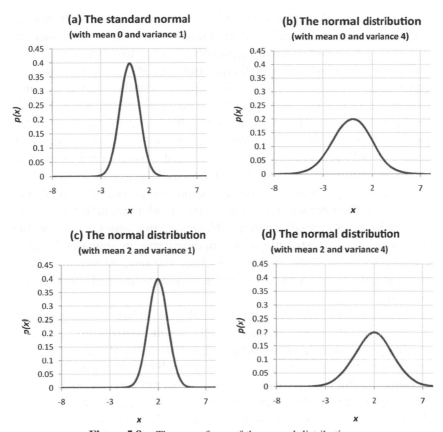

Figure 5.8.   The many faces of the normal distribution.

variance one. The standard normal distribution plays an important role in statistics, because it turns out that you can transform any normal random variable into a standard normal random variable. Simply subtract off the mean and divide by the standard deviation (the square root of the variance) and you're done. Mathematically, if X is a normal random variable with mean μ and variance σ^2, then

$$Z = \frac{X - \mu}{\sigma}$$

is a standard normal random variable.

Before the age of the Internet and Excel, transforming your normal random variable to a standard normal random variable was an essential step in the process of calculating probabilities for this distribution. This is because the cumulative distribution function, $P\{X \leq x\}$, for a normal random variable doesn't have a nice simple formula. Rather, we must rely on data tables and numerical approximations to calculate it. There's not enough paper in the world to print tables for every possible mean and variance combination, so the tables in the back of most statistics textbooks only report values for the standard normal distribution. And so it was up to you, the data analyst, to make the transformation.

Now, with programs like Excel and normal probability calculators available on the Internet, transforming your random variable to standard normal isn't so crucial anymore. However, this special case of the normal distribution still plays an important role in statistics because it's used for most basic analyses. So keep the random variable Z and the standard normal distribution in the back of your mind. We'll return to it later.

It's hard not to like the normal distribution. It has a common shape and nice, easy-to-work-with properties. But this distribution isn't just another pretty face. It's also incredibly useful. Not only do many data-sets follow this girl-next-door of all probability distributions, other random variables do as well. In fact, some random variables are so bell-shaped, the much easier normal distribution can be used in their place. The binomial and Poisson random variables are two such normal wannabes. I'll discuss the normal approximation to a binomial random variable here. The case for a Poisson random variable is similar and can be found in Triola (2004) and other references listed at the end of this book.

In general, the shape of the binomial distribution with parameters N and p depends on its mean, $\mu = Np$. When the mean is small, it has a right-skewed shape. However, as Np (and conversely $N(1 - p)$) grows, the shape gradually becomes more symmetric until eventually, it resembles a bell. Figure 5.9 shows how the shape of the binomial distribution changes for $N = 10$ when p changes from 0.1 to 0.5. A normal density function has been plotted over both cases for reference.

When $Np \geq 5$ and $N(1 - p) \geq 5$, the normal distribution provides a good approximation to the binomial distribution. To use this approximation for $P\{X \leq x\}$, you set the mean $\mu = Np$ and the variance $\sigma^2 = Np(1 - p)$. Then you make a *continuity correction*, a 0.5 change

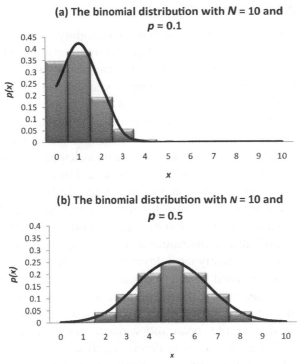

Figure 5.9. The binomial distribution: a normal wannabe.

to your x-value that adjusts for switching from a discrete distribution to a continuous distribution. This adjustment can be an addition or subtraction depending on the type of probability you want to calculate, but for both $P\{X \le x\}$ and $P\{X > x\}$, this means adding 0.5 to the x-value. Plug the mean, variance, and continuity-corrected x-value into a normal probability calculator like NORMDIST, and you're done.

Recall the friendly neighbor example from the previous section. The number of moderately chatty neighbors in a neighborhood with forty people is a binomial random variable with $N = 40$ and success probability $p = 0.16$. The mean number of moderately chatty neighbors is $Np = 40(0.16) = 6.4$, and the quantity $N(1 - p) = 33.6$. Both quantities are above five and so the normal approximation applies. The variance is $Np(1 - p) = 40(0.16)(0.84) = 5.4$, and so the standard deviation is $\sqrt{5.4} = 2.3$. To approximate $P\{X \le 10\}$, simply set $x = 10.5$, $\mu = 6.4$, and $\sigma = 2.3$, and run NORMDIST. The result is 0.96, very close to the value of 0.95 obtained from binomial distribution directly.

Sampling Distributions

Any time you collect data, you have uncertainty to deal with. This uncertainty comes from two places: (1) inherent variation in the values a random variable can take on and (2) the fact that for most studies, you can't capture the entire population and so you must rely on a sample to make your conclusions. I've spent most of this chapter showing how probability distributions can be used to describe the first type of uncertainty. As it turns out, they can also help us with the second.

A probability distribution for any value calculated from a dataset is called a *sample distribution*. This value could be an estimate such as the sample mean or standard deviation. It could also be a *statistic*, a formula that combines estimates together for reasons you'll see later. In any case, a sample distribution incorporates both types of uncertainty into a single probability distribution. For example, the sample mean and variance are commonly used descriptive statistics. These parameter estimates are often used to approximate the mean and variance of a population, in other words, the actual parameters, the values you don't know. Because of their importance in statistics, it's useful to know just how accurate the sample mean and variance are likely to be. Sample distributions give us the tools to determine this.

Sample distributions are the cornerstone of many statistical analyses, and with just a few of them in your toolbox, almost anything is possible. One in particular, based on the all-important central limit theorem, gives us yet another reason to love the normal distribution.

Fitting in with the Central Limit Theorem

The normal distribution works great for nicely rounded, symmetric datasets. It also works for nearly-normal data, those that are just slightly skewed or barely off center. But what about everybody else? What about those distributions with extreme values, heavy tails, and odd, chunky shapes? Well, the normal distribution helps us in this situation as well. There's a rule, and it's one everybody can appreciate, even the outcasts, nerds, and geeks. This rule can help virtually any dataset look normal. It's called the central limit theorem.

The *central limit theorem* applies to a sample distribution, specifically the distribution of the sample mean of a dataset. In words, the sample mean of any dataset, no matter how oddball, looks more and more normal as your number of observations grows. Here's the formal definition:

> As the sample size N increases, the sample distribution of the average (or sample mean) grows increasingly more like the normal distribution with (true) population mean μ and variance σ^2/N.

There are three important properties stated in the central limit theorem. To understand basic statistics, you need to know all three:

1. As you increase your sample size, the *sample mean* looks more and more normal. This is different from your individual observations looking more and more normal. No theorem can help you with that.
2. As you increase your sample size, the sample mean gets closer and closer to the true population mean (the value you're trying to estimate). This is one reason to love more data and not less.
3. The variance of the sample mean is σ^2/N. In other words, the variance of this estimate decreases proportionally with the number of observations in a sample. A smaller variance means more certainty in your conclusions. Yet another reason to love more data and not less.

The best way to illustrate the central limit theorem is to show it in action, and I'll do this for age data from the 2010 American Community Survey. Figure 5.10a plots the frequency distribution of the ages of respondents in the survey along with a scaled normal probability density function for reference. These data do not follow the popular bell-shaped distribution. They're chunky and square. There's no real peak in the center and no gradual slope to the sides. In other words, these data aren't normal.

According to the central limit theorem, it doesn't matter what the raw data look like, the sample variance should be proportional to the number of observations and if I have enough of them, the sample mean should be normal. To see if this is true, I've divided this large dataset of over 98,000 ages into what I'll call mini-samples, where each mini-sample contains the same number of randomly selected observations. If the sample means (averages) of these mini-samples grow more normal as the number in each mini-sample grows, then their frequency

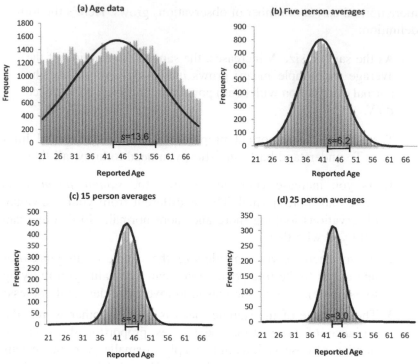

Figure 5.10.   Central limit theorem: the power of N.

distribution should start to have that nice, symmetric, bell shape. Figure 5.10b–d shows the frequency distribution of the mini-sample means, where each mini-sample has at first five observations (b), then 15 (c), then 25 (d). The sample standard deviation s is provided on each plot for reference.

The figure says it all. The standard deviation of the sample means gets smaller from (a) to (d), decreasing from 13.6 for the original observations to 3.0 for the 25 size sample means. The shape of the frequency distribution also changes, becoming more and more bell-like. By Figure 5.10d, the empirical distribution and the normal distribution are nearly indistinguishable. This is typical, for the sample mean of 25 or more observations to look approximately normal, and it gives us a rule of thumb for applying the normal distribution to a set of data. You'll see this rule of thumb again in several of the case studies that follow.

THE POWER OF *N*

There are many statistical techniques developed for data that follow the normal distribution. It is the prom queen of all distributions after all. But even if your data do not conform to the normal distribution, it doesn't matter. You can still use the abundance of simple statistical techniques that are based on it. As long as you have enough observations in your dataset, and as long as you're using a sample mean, your conclusions will be accurate.

This is the power of N.

So, if you're that normal person out there, the 44-year-old woman named Smith with a family, a good job, great neighbors, and no political opinions, descriptive statistics may be enough. If your data are perfectly bell-shaped and well-behaved, you may not need probability distributions. You may not need approximations, or the central limit theorem, or any other rules that help slightly awkward data feel perfectly normal. You may not need any of this. At least, not until your next birthday.

BIBLIOGRAPHY

U.S. Census Bureau. "About Us." http://www.census.gov/aboutus/, accessed October 3, 2012.

PART 2

What a Few Estimates and a Little Probability Can Do for You

Men Are Insensitive and Women Are Illogical. Really: Asserting Your Stereotypes with Confidence

Men love cars. Women can't drive.

Men love to watch sports. Women love to shop.

Men never stop and ask for directions. Women can't follow a map.

Men hate to talk about their feelings. Women love to talk on the phone.

Men are clueless when it comes to colors. Women are clueless when it comes to math.

Men fear commitment. Women fear being single.

Men are logical. Women are intuitive.

Stereotypes. Scientists have been studying them for years. In a recent search on Web of Science™, one of the Internet's largest repositories for scientific literature, thousands of articles about stereotypes

The Art of Data Analysis: How to Answer Almost Any Question Using Basic Statistics, First Edition. Kristin H. Jarman.

popped up. From misperceptions about religious people to misperceptions about female hockey players, the list of topics is endless. Many of these studies set out to disprove a stereotype, showing once and for all that people are the same, no matter what gender or hair color or sports preferences we have. Some of these studies succeed. Others show subtle differences. Still others claim these differences are not so subtle.

Maybe you have a favorite stereotype, a sweeping generalization you just can't let go. Are you convinced, for example, that all men are insensitive? Maybe you think women are horrible drivers. If you do, then prove it. It's the only way to assert yourself with confidence.

STATISTICIANS LOVE BIG DATASETS AND OTHER HALF-TRUTHS

In my twenty years as a statistical researcher, there's one stereotype I've run across hundreds of times, and it usually rears its ugly head within ten minutes of an initial consultation. It's the idea that statisticians love data, the more the better. That's not entirely true. What we really love is accurate conclusions, those that make us look smart, insightful, and maybe even a little clairvoyant. The only way to get accurate conclusions is with really good data, the kind that come from a well-designed study, the kind that truly represent the underlying phenomenon we're trying to characterize. More good data means more accurate conclusions. More bad data only means more work for us.

Take estimates, for example. Estimates are frequently used to make conclusions about a population. *Parameter estimates*, values that approximate some parameter of a population's probability distribution, are particularly important. The most common parameter estimates include the sample mean (an estimate of the population mean), the sample standard deviation (an estimate of the population standard deviation), and relative frequencies (an estimate of population proportions or probabilities). As good as any given estimate may be, it's still just a best guess constructed from a sample of data. It may be off by a little. It may be off by a lot. In other words, there's some uncertainty associated with any estimate. How much uncertainty? Confidence intervals help us answer this question.

A *confidence interval* is a range of numbers surrounding an estimate. This range measures how good your guess is, how much uncertainty is associated with it. A wide confidence interval suggests the estimate has a lot of room for error. A narrow confidence interval suggests the estimate is likely to be accurate. Say you have a dataset with sample mean 10. A confidence interval of 8.2 to 11.8 means you can be confident the true value, the population mean, falls between these two values. In other words, the margin of error is ±1.8 on either side of your sample mean. A confidence interval of 9.1 to 10.9 is half the width of this first interval, suggesting a much smaller margin of error, ±0.9 on either side of the sample mean. In both cases, the sample mean is the same, but the uncertainty associated with this estimate is very different.

Confidence intervals can be used not only to find the margin of error for an estimate, but also to compare a population to some preconceived notion of what it should be. For example, suppose you set out to prove that men fear commitment, your initial hypothesis being that 40% of all men suffer from this ailment. You and your single girlfriends carefully design a study, selecting men at random and confronting them with situations where their level of commitment-phobia can be measured. In the end, 30% of the men in your study fear commitment, with a confidence interval of 28 to 32%. If you designed your study properly and got good data, you could be confident that between 28 and 32% of all men suffer from commitment-phobia. Because the upper limit, 32%, is lower than your preconceived notion of 40%, you'd have to conclude that men do not fear commitment, at least, not as much as you previously thought. Stereotype busted.

Let's see how confidence intervals can be used to look at other gender stereotypes.

FINDING CONFIDENCE IN YOUR DATA

Most commonly used confidence intervals rely on a small number of sampling distributions. First introduced in the last chapter, a sampling distribution is the probability distribution of any estimate or statistic, in other words, a value calculated from a dataset. Like any probability distribution, the sampling distribution has its own mean and variance,

Statistic	Distribution	Caveats	Used for:
$Z = \sqrt{N}\dfrac{\bar{X} - \mu}{\sigma}$	Standard normal/Normal with mean 0 and variance 1	Sample comes from a normal distribution or $N>25$	Confidence intervals and hypothesis tests for the mean—variance known
$T = \sqrt{N}\dfrac{\bar{X} - \mu}{S}$	Student t-distribution with $N - 1$ degrees of freedom	Sample comes from a normal distribution or $N>25$	Confidence intervals and hypothesis tests for the mean—variance unknown
$Z = \dfrac{N\hat{p} - p}{\sqrt{N\hat{p}(1 - \hat{p})}}$	Standard normal/Normal with mean 0 and variance 1	$Np \geq 5$ and $N(1 - p) \geq 5$	Confidence intervals and hypothesis tests for a proportion

Figure 6.1. Common sample distributions.

both of which are used to construct a confidence interval. I'll introduce three sampling distributions in this chapter and show how they're used to construct common confidence intervals for a proportion and for the mean of a population. These three sampling distributions are so common throughout statistics, I've summarized them in the table provided in Figure 6.1.

Confidence Interval for the Mean of a Population

Suppose you're convinced all women love to talk on the phone and you'd like to prove it. You cajole every woman you know into handing over her cell phone bills and you log the number of minutes each one has used for the past three months. Even if you're the most outgoing male on the planet, you don't know every woman in the world. Your sample is just a subset of the entire population of women. And because you don't have the phone records of every woman on the planet, any estimates you calculate could be off by a little, or a lot. In order to assert your stereotype with confidence, you need a margin of error for the mean cell phone usage of women. In other words, you need a confidence interval.

There are two basic confidence intervals for the mean of the population, and the one you use depends on what you know going into the data analysis. Specifically, if you already know the variance of your population with certainty, you can use the simplest type of confidence interval, one that relies on concepts already presented in Chapter 5. On the other hand, if you don't know the variance of your population, you need to estimate it. In this case, you need a confidence interval that takes uncertainty in both the sample mean and sample standard deviation into account. I'll present both types of confidence intervals, starting with the simplest case first.

Known Variance

When you know the population variance, the confidence interval for the mean can be constructed from the following sampling distribution:

Probability distribution of a z-statistic: For a set of observations x_1, x_2,   , x_N with sample mean $\bar{x}$ that follows a normal distribution, unknown population mean μ, and known variance σ^2, the value

$$Z = \sqrt{N}\,\frac{\bar{x} - \mu}{\sigma}$$

follows a standard normal distribution, in other words a normal distribution with mean 0 and variance 1.

The z-statistic is what I think of as a statistical difference, in other words, the difference between the sample mean and the true (unknown) population mean, scaled to take variation into account. This scaling process transforms the difference to a standard normal distribution, allowing us to evaluate its size on a common scale.

This result plus a little algebra is all we need to construct a confidence interval, and the process starts with an interval probability. An *interval probability* is the probability a random variable will fall within some range, between a lower value, a, and an upper value, b. Illustrated in Figure 6.2, this probability represents the shaded area underneath the probability distribution function between the two limits. Interval probabilities can be calculated by adding up (or integrating) the probability distribution (or density) function for all values between a and b. More

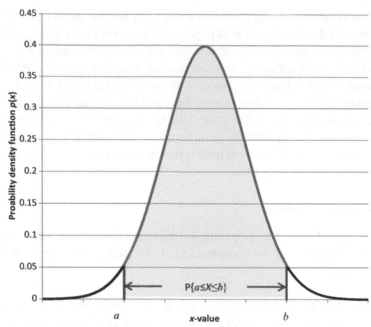

Figure 6.2. The interval probability.

commonly, though, they're calculated using the cumulative distribution function. Recall, the cumulative distribution function, or CDF, is the probability $P\{X \leq x\}$. As suggested in Figure 6.2, the interval probability $P\{a \leq X \leq b\}$ is just the cumulative distribution $P\{X \leq b\}$ with the region where $X \leq a$ ignored. This leads to the following useful result:

The interval probability $P\{a \leq X \leq b\}$ is the same as

$$P\{a \leq X \leq b\} = P\{X \leq b\} - P\{X \leq a\}.$$

How do interval probabilities apply to confidence intervals? Suppose you'd like to find a range $-z_{crit}$ to z_{crit} so that the probability the z-statistic falls inside this range is high, say 0.95. This is the following interval probability:

$$P\{-z_{crit} \leq Z \leq z_{crit}\} = 0.95.$$

The z-statistic is defined by

$$Z = \sqrt{N} \frac{\bar{x} - \mu}{\sigma},$$

so you can make this substitution inside the interval probability and do a little algebra to get

$$P\left\{ \bar{x} - z_{crit}\, \frac{\sigma}{\sqrt{N}} \le \mu \le \bar{x} + z_{crit}\, \frac{\sigma}{\sqrt{N}} \right\} = 0.95.$$

This range,

$$\bar{x} - z_{crit}\, \frac{\sigma}{\sqrt{N}} \le \mu \le \bar{x} + z_{crit}\, \frac{\sigma}{\sqrt{N}},$$

is the 95% confidence interval for the mean μ when the variance is known. Note the width of this confidence interval depends on the population standard deviation, σ, and the number of observations in your sample, N. You'll find this to be a recurring theme in statistics, that the error or uncertainty of a result depends on the variation scaled by the sample size. This is because the variance of an estimate, such as the sample mean, decreases proportionally with the number of samples in a data set. In other words, according to the central limit theorem from last chapter, more observations are always better.

There's one other value that affects the width of the interval, z_{crit}. This is the *critical value*, the number that makes the interval probability $P\{-z_{crit} \le Z \le z_{crit}\} = 0.95$ true. Figure 6.3 illustrates the process of calculating a critical value. The goal is to find z_{crit} so that the area between $-z_{crit}$ and z_{crit} is the specified probability, 0.95 in this case. This means the leftovers, the *tail probabilities*, must add to $1 - 0.95 = 0.05$, a value denoted by the Greek letter α. This leftover probability α is often called the *significance level*, and each of the two tails make up half of this significance level, or $\alpha/2 = 0.025$. Because two tail probabilities—a left tail and a right tail—are included in the probability calculation, I'll call this a two-tailed critical value. This is to distinguish it from the one-tailed critical value that will be discussed at length in the next chapter.

The left-hand tail in Figure 6.3 is just the cumulative probability $P\{Z \le -z_{crit}\}$, so the two-tailed critical value z_{crit} can be calculated as the value for which $P\{Z \le -z_{crit}\} = \alpha/2 = 0.025$. Most statistics texts have comprehensive tables of critical values for every significance level you could want, and most of these tables report the values to several decimal places. In the real world, however, only a few significance levels are commonly used and you rarely have a situation where it matters if the critical value is 1.64 or 1.65. So I've simplified the typical

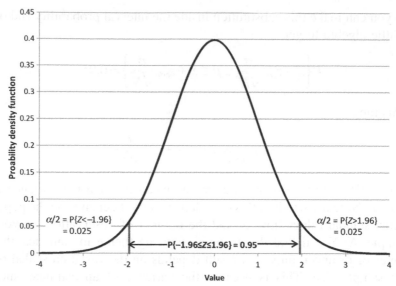

Figure 6.3.   Two-tailed critical values for the standard normal distribution with a = 0.05.

table and provided critical values to one decimal place in Appendix A, Figure b for easy reference. This figure already takes both tails into account and so no adjustment to α is necessary. As an example, for a 95% confidence interval, the two-tailed critical value is $z_{crit} = 2.0$.

Suppose you have the phone bills of 20 women and you know the standard deviation to be $\sigma = 35$ minutes per week. If the sample mean (average) is $\bar{x} = 198$ minutes per week, the 95% confidence interval for the population mean is

$$\left(198 - 1.96 \frac{35}{\sqrt{20}}, 198 + 1.96 \frac{35}{\sqrt{20}}\right), \quad \text{or} \quad 182 - 213.$$

This range is the margin of error on your sample mean.

It's natural to take a statement like this and assume there's a probability of 0.95 the true mean falls between 182 and 213. But this isn't the case. The true mean is some fixed constant value μ, even if you don't know what it is. The uncertainty in the confidence interval comes from the sample estimate $\bar{x}$, not μ. So rather than making claims about the probability of the population mean μ, you'd instead say something like, "I'm 95% confident the mean cell phone usage for women is between 182 and 213 minutes per week." It's a subtle but important distinction.

Unknown Variance

The confidence interval for a mean with known population variance is the simplest to construct and the easiest to understand. Unfortunately, in my years as a statistician, I've only run across this situation a few times. Usually, when you don't know the mean, you don't know the variance either. You need to estimate both. In this case, a z-statistic no longer applies. But there is a common distribution that does apply, the *Student's t-distribution*.

Probability distribution of the t-statistic: for a set of observations $x_1, x_2, \ldots, x_N$ with average $\overline{x}$, sample standard deviation s, unknown mean μ, and unknown variance σ^2, the value

$$T = \sqrt{N}\, \frac{\overline{x} - \mu}{s}$$

has Student's t-distribution with $N - 1$ degrees of freedom.

The t-statistic should look familiar. It's basically the z-statistic with the sample standard deviation s replacing the population standard deviation σ. Like the z-statistic, the t-statistic is a statistical difference, scaled to account for the variation in the sample. Like the standard normal distribution, the t-distribution has a mean of zero and a symmetric, bell-like shape. But these two distributions are not the same. Substituting the sample standard deviation for the population standard deviation adds uncertainty to the statistic, and so the t-distribution has a wider bell shape than its normal counterpart.

One of the ways in which the t-distribution accounts for uncertainty in the sample is through the degrees of freedom parameter. Remember, most statistical methods rely on an independent sample, a group of observations that are independent of one another and representative of the population as a whole. The *degrees of freedom* term measures the number of independent values going into a test statistic or estimate; in other words, the net sample size. When you start with N data values, you have N degrees of freedom. However, you lose one degree of freedom for substituting the sample standard deviation for the population standard deviation. This gives a net number of independent values $N - 1$.

A confidence interval for the mean with unknown variance can be calculated just like it was for known variance. You simply set up an interval probability for the t-statistic and do a little algebra. There are just two differences: the substitution of the sample standard deviation s in place of σ, and $t_{crit, N-1}$, the $\alpha/2$ critical value for the t-distribution, in place of z_{crit}. This process is instructive and if you're comfortable with interval probabilities and algebra, I urge you to try it. You should get the following interval:

$$\left(\bar{x} - t_{\frac{\alpha}{2}, N-1} \frac{s}{\sqrt{N}}, \bar{x} + t_{\frac{\alpha}{2}, N-1} \frac{s}{\sqrt{N}} \right).$$

The critical value for the t-distribution with $N - 1$ degrees of freedom can be calculated in Excel using TINV. Most statistics texts also provide comprehensive tables of critical values for this distribution. I've simplified these tables in Appendix B, which provides two-tailed critical values of the t-distribution for the most common confidence levels and a handful of sample sizes. For example, on a 95% confidence interval for the mean, the critical value for a sample with $N - 1 = 20$ degrees of freedom is 2.1. If you have a number of degrees of freedom that do not appear in the table, then you can approximate the critical value by averaging the two surrounding values. For a 95% confidence interval, 15 degrees of freedom is halfway between 10 and 20 degrees of freedom, and so the critical value for this case can be approximated by $(2.2 + 2.1)/2 = 2.15$.

Back to estimating women's weekly cell phone usage. Suppose you don't know the variance of the number of minutes a woman talks on her phone every week. You need to estimate it from the data. Suppose it comes to $s = 35$. To construct a confidence interval for the mean, you'd use the sample mean 198, *sample* standard deviation $s = 35$, and $N = 20$ in the above confidence interval. The $\alpha = 0.05$ critical value of the t-distribution with 19 degrees of freedom is 2.1. Therefore, the 95% confidence interval for μ is 181,214.

You might notice the critical value for Student's t-distribution with 19 degrees of freedom (2.1) is larger than the corresponding value for the standard normal distribution (1.96). This larger critical value makes for a wider confidence interval in the case where the variance is not known. This is always the case, that a confidence interval for the mean with unknown variance is larger than the corresponding interval for known variance. The increased margin of error is the price you pay for not knowing the variance and being forced to estimate it from a dataset.

Confidence Interval for a Proportion

Maybe you think all men love to watch sports on TV. To prove this, you make an official-looking log sheet and give it to every guy you know so they can record how many games, matches, and contests they watch in a single weekend. With these data, you could estimate the proportion of the men in your sample who watch four, five, or even ten games in those two days. But it's not good enough to prove most men spend their weekend on the couch. To state your case with confidence, you need to put a margin of error on your estimate.

Like a confidence interval for the mean, the confidence interval for a proportion (or probability) starts with the sampling distribution for this estimate. You may remember from the last chapter that the number of successes in N independent trials is a binomial random variable, in other words, it can be described by the binomial distribution. This same distribution can be used to describe the number of game-watching couch potatoes. In particular, suppose you have 35 men in your study, and you define a sports-lover as a man who watches at least three games in a weekend. If your sample is a random sample, then the number of sports-lovers follows the binomial distribution with $N = 35$ men and unknown probability p that any one man is into sports.

To construct a confidence interval around p, you could start with the binomial distribution and an interval probability and perform the required algebraic manipulations, but the binomial distribution is difficult to work with. Instead, most people use the normal approximation to the binomial distribution. This approximation was presented in the last chapter and gives the following sampling distribution.

Probability distribution of a proportion: If $\hat{p}$ is the proportion of successes in a sample having N observations, and if this sample represents a population with true (unknown) success probability p, then when $N\hat{p} \geq 5$ and $N(1-\hat{p}) \geq 5$, the statistic

$$Z = \frac{N\hat{p} - p}{\sqrt{N\hat{p}(1-\hat{p})}}$$

can be approximated by the standard normal distribution.

From here, calculating the confidence interval follows exactly the same path as the confidence interval for a population mean, and once again, if you're mathematically inclined, I urge you to try it for yourself. A little algebra gives the following confidence interval for a population proportion:

$$\left(\hat{p} - z_{\frac{\alpha}{2}} \sqrt{\frac{\hat{p}(1-\hat{p})}{N}}, \; \hat{p} + z_{\frac{\alpha}{2}} \sqrt{\frac{\hat{p}(1-\hat{p})}{N}} \right).$$

Like the confidence interval for the population mean with known variance, the critical value here can be calculated from the NORMINV function in Excel or by referring to the table of two-tailed critical z-values in Appendix A.

Suppose you give your sports log sheet to 35 men, and each one returns the sheet complete with a tally of the number of games he watched over a particular weekend. You calculate the proportion of men who watched at least three games, and you get 42%, or $\hat{p} = 0.42$. For a 95% confidence interval, the critical value is $z_{0.025} = 1.96$ and so the 95% confidence interval on the proportion of men who watch at least three games per weekend is

$$\hat{p} = 0.42 \pm 1.96 \sqrt{\frac{0.42(0.78)}{35}}$$
$$= 0.42 \pm 0.19.$$

This gives a range of 0.23 to 0.61. In other words, you can be 95% confident that the number of weekend sports-watching warriors is between 23 and 61%.

Confidence intervals are often reported in this way, as a sample proportion (or mean) plus or minus some margin of error. This form is convenient. It simplifies the mathematical expression and emphasizes the margin of error. Figure 6.4 presents confidence intervals for the mean and proportion in this way for easy reference.

ARE MEN INSENSITIVE? ARE WOMEN ILLOGICAL?

There are probably lots of highly scientific studies that have tried to prove or disprove these stereotypes, studies rooted in psychology, and

Confidence interval	Formula	Where to find the critical value
Population mean, variance known	$\bar{x} \pm Z_{crit}\dfrac{\sigma}{\sqrt{N}}$	Appendix A
Population mean, variance unknown	$\bar{x} \pm t_{\frac{\alpha}{2}, N-1}\dfrac{S}{\sqrt{N}}$	Appendix B
Population proportion	$\hat{p} \pm Z_{\frac{\alpha}{2}}\sqrt{\dfrac{\hat{p}(1-\hat{p})}{N}}$	Appendix A

Figure 6.4. Common confidence intervals.

taking into account hormones, anatomy, and brain chemistry. This isn't one of those studies.

I have no idea how to measure whether the logic neurons are different between men and women. Nor do I have any idea how to measure activity in the sensitivity center of a person's brain. In fact, I don't even know if there is such a thing as a logic neuron or a sensitivity center. No, this study is anything but a highly scientific, controlled experiment. It doesn't even directly measure men's sensitivity and women's logical thinking. In other words, any conclusions you draw from this chapter will be drawn at your own risk.

Since I can't measure sensitivity and logical thought directly, I'll scale down my questions accordingly. In my experience, men often complain that women are illogical simply because women appear to be more emotional. Likewise, women often complain that men are insensitive because they seem to be more interested in facts and figures than human emotions. If this is true, then the driving force behind these two stereotypes has more to do with emotional reactions than a person's capacity for sensitivity or logical thought. This is something I can measure. But rather than making complex arguments based on psychology and neuroscience, I'll simply ask people what they're thinking (or feeling, as the case may be).

This is apparently what Isabel Myers was thinking in the mid-1900s when she proposed a simple method for typing personalities. She argued that a person's behavior depends on what he or she is thinking. So why not just present a person with situations and ask how he or she would handle it? This is what she did, and with her mother, Katherine Briggs, she created one of the most widely used personality tests of any ever developed.

The Myers-Briggs personality test (http://www.myersbriggs.org/) is a simple test with seventy multiple choice questions, each question having two possible answers. It doesn't measure intelligence, or psychological health. It only characterizes how a person thinks about and reacts to the world. The test places people into categories based on four characteristics, each of these four characteristics represented by a letter. The first characteristic, represented by an E or I, indicates whether you are extroverted or introverted. An extrovert draws energy from socializing with others, while an introvert draws energy from being alone. The second characteristic, indicated by an S or N, indicates whether you more sensing or intuitive. A sensing person tends to focus on the outside world, and an intuitive person tends to add his or her own internal meaning and interpretation to a situation. The third characteristic, represented by T or F, refers to thinking or feeling. A thinking person tends to focus more on facts and logic, while a feeling person tends to focus more on harmony and the feelings of others. The last characteristic, indicated by a J or P, refers to judging or perceiving. A judging person likes to have things decided upon and scheduled, while a perceiving person likes to keep his or her options open.

These eight letters combine into 16 different personality types, the idea being that people of the same type share certain characteristics. For example, the letters ESTJ (extroverted-sensing-thinking-judging) refer to a personality type known as the Supervisor. Like the name implies, Supervisors have a tendency to take charge of a situation, making sure everybody toes the line, giving direction as needed. They're also down to earth, reliable, and willing to donate their time and energy to their favorite organizations. You probably know a few Supervisors. Most people do. It's estimated they make up about 10% of the population (Keirsey 1998).

If men are more interested in facts, and women more interested in feelings, these tendencies should show up in the results of the Myers-Briggs personality test, specifically through the answers to the thinking versus feeling (T vs. F) questions. There are twenty such questions on the test (Keirsey 1998), and if these stereotypes hold, men will tend to be what I'll call T people, those who pick T answers more often than F answers. Conversely, women will tend to be F people, those who pick F answers most often.

The version of the Myers-Briggs test printed in *Please Understand Me* (Keirsey 1998) clearly indicates which questions go with which characteristics, and a written test, as opposed to one of the many online tests, gives me access to the number of T versus F answers. So my experimental plan was to get men and women to take this test, estimate the mean number of T answers for men and women, and put confidence bounds on the result. Because the sample mean of 25 or more observations can generally be approximated by the normal distribution (see the last chapter for a discussion of this), my goal was to get at least 50 tests, 25 from men and 25 from women.

Because any statistical analysis is only as good as the data you feed into it, it's important to collect a sample that represents the population as a whole. For example, suppose you're estimating the unwillingness of men to stop and ask for directions. And suppose you only collect data from your five older brothers. Your estimate and your confidence interval could be skewed by the fact that you come from a long line of very stubborn people. In other words, based on a few hard-headed men, your brothers, you might falsely conclude that all men refuse to ask for directions. Ever. To prevent such false stereotypes from taking root in your brain and coming back to bite you later, you need a proper sample to work with.

The best type of sample for my study would be a random sample. And this is what I gathered. Well, sort of. I actually carried around a copy of the test with me for a few weeks and begged virtually every adult I spoke with to take it. (I stayed away from work, however, because most of my colleagues are statisticians or scientists and too many of them could really bias my sample). People weren't always thrilled about sharing this kind of information with a strange, portfolio-wielding woman. Some might have even found me a little creepy. But almost everybody cooperated with me, and in the end, I got 51 completed tests. Thanks to all the family, friends, acquaintances, and total strangers who participated.

Because the group included people I knew and strangers I happened across, this isn't a classic random sample. So, I also asked all of the participants their age and occupation, just to make sure I wasn't getting an overabundance of, say, 37-year-old accountants in my dataset. Fortunately, I didn't. My test subjects were between the ages of 19 and 70. Among them were homemakers, students, retirees, doctors, scientists, businessmen, teachers, a pipe fitter, an editor, and even a city councilman.

It's always a good idea to do what I call a sanity check, just to make sure your data are consistent with what you expect, but it's especially important when you aren't working with a pure random sample. I addition to checking the age and occupations of my volunteers, I also wanted to make sure the overall proportion of T people and F people is consistent with the population as a whole. Of the 51 people in my study, 23 (45%) were T people, 23 (45%) were F people, and five (10%) answered equal numbers of both. With 51 test subjects and a relative frequency of 0.45, the 95% confidence interval on the proportion of T people is

$$0.45 \pm 1.96 \sqrt{\frac{0.45(0.55)}{51}} = 0.45 \pm 0.14.$$

In other words, according to my sample, the proportion of T people, those who gave mostly T answers to the T versus F questions, is between 0.31 and 0.59, with a best estimate of 0.45. The Myers-Briggs Foundation reports a population percentage of T people to be 40.2%, a proportion of 0.40 (Myers-Briggs Foundation 2012). Since this population proportion falls within the margin of error for my study, I can claim they're consistent with one another.

The agreement between my result and the Myers-Briggs Foundation numbers gives me a good feeling about the data I've collected, but it doesn't answer the question at hand. To determine if men tend to be T people and women tend to be F people, I need to split my sample into two groups: men and women. If the proportion of men who are T people is higher than 0.40, the population value, the stereotype about men being fact-focused is supported. Likewise, if the proportion of women who are T people is lower than 0.40, then the stereotype about women being emotional is supported.

Out of 27 women in the study, 19% tested as T people. The 95% confidence interval for the proportion 0.19 is (0.04, 0.33). The upper limit on this confidence interval is below the total population proportion, suggesting women do tend to be F people. Out of 24 men in the study, 75% tested as T people. The 95% confidence interval on this

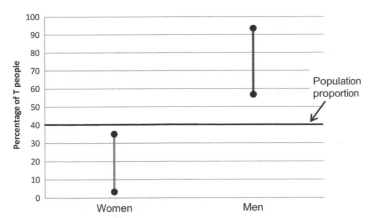

Figure 6.5.   Confidence intervals for the proportion of T people.

proportion, 0.75 is (0.58,0.92). The lower limit on this interval is above the population proportion, suggesting men are more likely to be T people. These results are illustrated in Figure 6.5.

This analysis may support the stereotypes, but it doesn't tell the whole story. On this personality test, a person is categorized into the thinking or feeling group based on a majority of answers to 20 questions. A person who gives T answers to all twenty questions is categorized the same as a person who gives T answers to 11 questions, even though the reliance on facts of these two individuals may be completely different. So, in addition to looking at the proportion of T people, it might be useful to look at the number of T answers for men and women to see how much they differ from ten, the midpoint suggesting equal T and F tendencies. Figure 6.6 plots the frequency distribution of the number of T answers for men and women. The sample mean and standard deviation are added for reference.

Figure 6.6 shows that while the average number of T answers is higher for men than women, the two frequency distributions overlap quite a bit and both of them straddle the midpoint of 10. Men vary in their number of T answers from eight to 19, a range of 11. Women vary even more, ranging from zero all the way up to 15. This means that while men tend to be T people and women tend to be F people, there are plenty of people who don't fit this profile. Let's see how these results translate into confidence intervals for the mean.

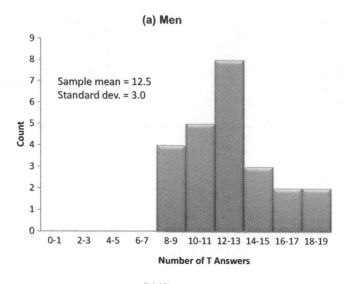

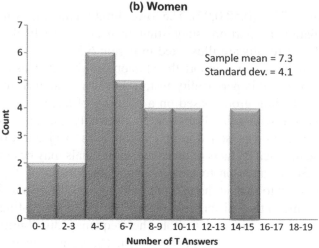

Figure 6.6.   Frequency distribution for number of T answers.

For these data, the variance of the population is unknown and so the 95% confidence interval for the mean uses the "variance unknown" formula in Figure 6.4. For men, the sample mean is 12.5, the sample standard deviation is 3.0, the number of observations is $N = 24$, and the $\alpha = 0.05$ two-tailed critical value $t_{\text{crit, }23} = 2.1$. This gives a confidence interval of

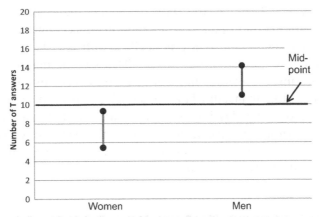

Figure 6.7. Confidence intervals for the number of T answers.

$$\mu = 12.5 \pm 2.1 \frac{3.0}{\sqrt{24}} = 12.5 \pm 1.25.$$

In other words, with 95% confidence, the true mean number of T answers for men is between 11.3 and 13.8. The lower confidence bound is above ten, suggesting men really do tend to be more T-like. But with a lower limit of just over ten T questions, it isn't by very much. For women, the sample mean is 7.3, the standard deviation is 4.1, $N = 27$, and the $\alpha = 0.05$ two-tailed critical value for the t-distribution with $N - 1 = 26$ degrees of freedom is $t_{crit, 23} = 2.1$, giving a confidence interval of 5.7 to 9.0. The upper limit is just below 10, suggesting women are more F-like. But again, not by much. These results are illustrated in Figure 6.7.

Figure 6.7 shows the average man is more T-like and the average woman more F-like. But does this mean men are more T-like than women? In other words, is the difference between the two sexes *statistically significant*, different than you'd expect based on random variation alone? Confidence intervals can be used to compare populations to one another, but this is not what they're meant to do and so the comparison must be done carefully. When the confidence intervals of two populations do not overlap, then we usually say there is a significant difference. When they do overlap, the difference may or may not be significant, and unless you can apply some more sophisticated statistical reasoning to the data, you need a direct comparison between the two groups to be sure. This is the subject of the next chapter.

As Figure 6.7 shows, the 95% confidence interval for men and women do not overlap, and so the average difference between the sexes is statistically significant. Is it practically significant? Is the difference really enough to validate a stereotype? That's the important question. For both groups, the average number of T answers is close the midpoint value of ten. More than a third of women were either T people or within one question of being classified so. More than a third of men were either F people or within one question of earning that label. In other words, while there does appear to be a real difference, there are still plenty of T women and plenty of F men.

THE BATTLE OF THE SEXES RAGES ON

How you interpret these results is up to you. You could argue whether this study proves men are more logical than women or simply more insensitive. You could argue whether it proves women are more empathetic than men or just more irrational. You could also argue that an average difference of just a few T questions on a personality test says nothing at all about men and women, particularly when there are so many men and women who don't fit the average profile. However you read them, these results can only add ammunition to the battle of the sexes, giving you one more way to assert your stereotypes with confidence.

BIBLIOGRAPHY

ABC News. "Cell Phones and Demography: Nielson Data Breaks Down Mobile Phone Usage by Race, Age, Gender, and Region." http://abcnews.go.com/WN/cell-phones-demography-nielsen-data-breaks-mobile-phone/story?id=11468925, accessed September 1, 2012.

Keirsey, David. 1998. *Please Understand Me*. Prometheus Nemesis.

Myers-Briggs Foundation. "How Frequent Is My Type?" http://www.myersbriggs.org/my-mbti-personality-type/my-mbti-results/how-frequent-is-my-type.asp, accessed September 1, 2012.

Myers-Briggs Foundation. "MTBI Basics." http://www.myersbriggs.org/my-mbti-personality-type/mbti-basics/, accessed October 3, 2012.

Schauer, Frederick. 2003. *Profiles, Probabilities, and Stereotypes*. Belknap Press.

CHAPTER 7

Godzilla versus King Kong: How Hypothesis Tests Can Be Used to Settle the Battle of the Movie Monsters

Picture this: Tokyo in ruins. Godzilla's back, and he's on the rampage. Buildings fall and cars crumple under the force of the giant lizard. Japanese mobs flee in panic. Suddenly, across the smoky skyline, a rustle disturbs the trees of a distant forest. King Kong, the 50-foot gorilla, emerges, and he's angry. But before he can crush his first victim, the two monsters spot one another. Godzilla screeches and flails his arms. King Kong roars and beats his chest. An epic battle ensues.

In the 1962 classic film *King Kong vs. Godzilla*, the monsters fight a long, drawn-out battle, gradually moving toward the beach until they both disappear into the ocean. Eventually, Kong is spotted swimming back to his home island. Godzilla isn't seen again. It appears the ape has beaten the lizard. But a monster's success isn't only measured by his destructive power. In this age of big movie budgets and bottom lines, popularity is the real king. Without lasting appeal, even the most

The Art of Data Analysis: How to Answer Almost Any Question Using Basic Statistics,
First Edition. Kristin H. Jarman.
© 2013 John Wiley & Sons, Inc. Published 2013 by John Wiley & Sons, Inc.

terrifying freak of nature won't live to terrorize Tokyo another day. So, who's more popular, King Kong or Godzilla? In this chapter, hypothesis tests will be used to settle the battle of the movie monsters.

HYPOTHESIS TESTS: LET THE JUDGING BEGIN

Hypothesis tests are the Supreme Court judges of statistics. They take evidence in the form of a sample of data and make judgments about some population. Every hypothesis test has three parts to it: (1) two competing claims about a population, (2) a test statistic that mathematically weighs the two claims against one another, and (3) a decision criterion, or rule for accepting one of the two claims and rejecting the other.

All hypothesis tests start with two competing claims, or hypotheses. The *null hypothesis*, or H_0, is a default statement about a population, the claim of innocence, if you will. The *alternate hypothesis*, H_A, is the competing statement, the guilty claim. The purpose of a hypothesis test is to determine if there's enough evidence to reject H_0 in favor of H_A, in other words, to prove guilt beyond a reasonable doubt. For example, in the hypothesis test

H_0: $\mu = 0$ vs.

H_A: $\mu = 1$,

the population mean is what's being judged. The null hypothesis claims it is zero. The alternate hypothesis claims it is one. The hypothesis test for these two competing claims would look at the data and determine if there's enough evidence to reject the default claim that the population mean is zero, thereby accepting the claim that it is one.

Like judges in a courtroom, every hypothesis test needs evidence on which to base its conclusions. A *test statistic* is just such evidence. It's a statistic, a value calculated from a sample, used to weigh the likelihood of one hypothesis over the other. Typically, the test statistic is a combination of a sample estimate scaled by its associated variation. For example, the test statistic for the hypothesis H_0: $\mu = 0$ versus H_A: $\mu = 1$ is a function of the sample mean, the standard deviation, and the number of observations in a sample.

The *decision threshold* provides the rule for judging your hypotheses. The evidence, the test statistic, is compared to this decision threshold. If it falls on one side of this threshold, you accept the null hypothesis H_0. If it falls on the other, you reject it in favor of H_A. There are two ways to express the decision threshold, one based on a critical value for the test statistic and the other based on a tail probability for the test statistic. As you'll see later in the chapter, these two different thresholds always lead to the same conclusion, and so one of them isn't necessarily better than the other. It's simply a matter of how you prefer to express your results.

There are hypothesis tests for just about any population parameter you can imagine—mean, variance, proportion—and while the details of each are different, all of them use a test statistic and a decision criterion to pit two competing hypotheses against one another. In this chapter, I'll introduce three of the most commonly used hypothesis tests. For others, I refer you to one of the textbooks listed at the end of this book.

TWO GIANT MONSTERS ON ONE SMALL ISLAND

King Kong first appeared on the big screen in the 1933 classic horror film named after him. In the movie, this giant, apelike creature is discovered living on remote Skull Island, a brutal place where he must defend himself against living dinosaurs and other atrocities, and where the islanders must sacrifice virgins to the fifty-foot beast in order to keep his rage from boiling over. All this changes when a group of modern explorers discover the monster. Using one of their own—a young actress named Ann Darrow—for bait, the explorers capture the beast and transport him to New York City where they put him on display for the world to see. But there's a wrinkle in this plan. Kong has developed a serious crush on Ann Darrow, and betrayal only adds to his rather significant anger issues. He breaks free from confinement and releases his rage on the Big Apple, storming the city in search of his crush and causing destruction in the process. Eventually, the beast ends up on top of the Empire State Building, clutching his love interest in one hand and fending off a swarm of helicopters with the other. In the end, Kong falls off the building, landing in a pile of lifeless fangs and fur.

Some twenty years younger than Kong, Godzilla first appeared in the 1954 movie sharing the lizard's name. The Japanese authorities grow concerned when ships begin exploding off the coast of their small country. A research party is sent to investigate. They learn a giant lizard has been wreaking havoc on nearby Odo Island, and when they visit, they get their first glimpse of the creature. The research party barely has time to alert the authorities before Godzilla heads for Tokyo and goes on the rampage, with no apparent motive other than the complete destruction of all things Japanese. Nothing seems to stop the monster, not fire, not electricity. Ultimately, the only thing that defeats Godzilla is a strange bomb-like device planted in the harbor by the scientist who developed it. The giant lizard settles down for an underwater nap and the scientist detonates the device, sacrificing his own life in order to kill the beast.

Of course, death cannot stop the likes of Kong and Godzilla. Both monsters became legends the moment they first appeared. Between the two of them, they've starred in countless movies, television shows, comic books, and video games. They've been the inspiration for action figures, cartoons, and just about any type of merchandising rip-off you can imagine. And there are literally thousands of websites that mention these two creatures, many of them touting the awesomeness of one over the other.

So does Godzilla or King Kong make the better movie monster? I'll come right out and admit that for me, it's Kong. Here's why:

- Kong came first.
- Kong was born on a tropical island. Godzilla was born in a nuclear explosion.
- Kong only has morning breath. Godzilla has atomic breath.
- Kong wears a thick fur coat. Godzilla wears scales.
- Kong has empathy and near human intelligence. Godzilla has scales.
- In the end, Kong gets the girl. Godzilla only gets more scales.

Of course, this is just one person's opinion. If I really want to prove Kong is the better monster, I need an unbiased way to compare the two of them, something that takes into account more than just my own preferences. That's why I'm going directly to the biggest repository of opinions in the world. The Internet. And I'll be using one of the most ubiquitous data sources out there. The top ten list.

Don't get me wrong, I'm not making a top ten list. That's already been done, over and over again. A recent Google search for "top ten movie monsters" resulted in over 11,000,000 hits. Not all of those web pages have what I need, some of them don't have lists at all, some point to others' lists, and still others focus on modern slasher movie villains like Freddy Krueger and Jason. But even if 99% of those links are useless for comparing Godzilla to King Kong, that still gives me over 100,000 top ten classic movie monster lists made by as many different monster movie fans. That's a good sample by anyone's definition.

So who makes these lists? Television channels, newspapers, special interest publications, film critics, bloggers, and random movie fans. Some are polls of readers or viewers, but most are simply the opinion of one person. By taking a sample of these many top ten lists, I can reach beyond the opinion of one individual and make conclusions that apply to a whole world of Internet users.

This was my data collection plan: to randomly select at least sixty top ten classic movie monster lists from the Internet—you'll see why sixty a little later—and compare the rankings for Godzilla and King Kong across these lists. The creature that ranks higher wins.

GODZILLA VERSUS KONG

After a couple hours of searching, I had Godzilla and Kong rankings from sixty top ten classic movie monster lists posted from as many different sources. From a quick look at the data, it isn't obvious who wins. For example, ReelzChannel TV recently conducted an online poll of its viewers and came up with a top ten list. Godzilla ranked number six. Kong didn't make the list. BBC News also has a top ten ranking, this one put together by a film critic. King Kong took the #1 spot on this list. Godzilla didn't appear anywhere. With this kind of variation in the rankings, it'll take more than a quick look at the data to pick a winner.

Before diving into the statistical analysis, I need to point out one important issue with these data. Among the top ten lists I found, some came from a poll of many people. Other lists were simply the opinion of one film critic or movie fan. Both types of lists rank the monsters from one to ten (or higher), and on the surface, it might seem they're perfectly compatible. But after after the last few chapters, I hope you're beginning to appreciate the subtle but important difference between a single data value, one person's opinion, and an estimate calculated from lots of data values, the results of a poll. The central limit theorem from

Chapter 4 tells us that the variance of an estimate decreases as the number of samples increases. This means the variance of the rankings from a poll is smaller than the variance of the rankings put together in one person's head. In other words, the two types of lists have different statistical properties.

This is a problem. Virtually all basic data analysis methods assume the individual observations have the same mean, variance, and probability distribution. The analysis I'm about to present is no different. I could ignore this assumption and go forward, combining the two types of lists without regard to how they were constructed. But if there are enough of these differences between observations, any estimate of the population variance will be inaccurate and so will my conclusions. Besides, this problem has an easy fix. If I simply remove the small number of polls in my dataset—there were only three—all my rankings will be from individuals and so every observation will have the same basic statistical properties. This is what I did.

Goodness-of-Fit Tests

Figure 7.1 plots the frequency distribution of the 57 remaining rankings for Godzilla (a) and King Kong (b). In the figure, the bin labeled ">10" represents the fraction of lists that didn't include the monster. Given the enduring popularity of both creatures, it's no surprise that both frequency distributions are concentrated around the top three spots. But is this concentration of #1–#3 rankings statistically significant, in other words, more than you'd expect by random variation alone? A goodness-of-fit test can help us find out.

A *goodness-of-fit test* is a hypothesis test for judging whether or not a frequency distribution conforms to some (theoretical) probability distribution. In general, if you have a population with M categories, the two competing hypotheses for a goodness-of-fit test are

H_0: $p_1 = q_1, p_2 = q_2, \ldots, p_M = q_M$ vs.

H_A: At least one p_k is not equal to its corresponding q_k.

where the p_k values are the relative frequencies for the population you're testing, and the q_k values are the corresponding theoretical probability values.

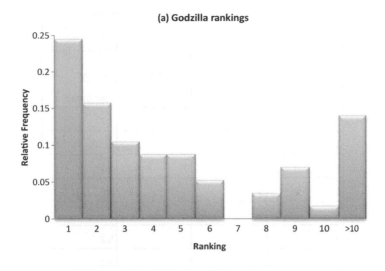

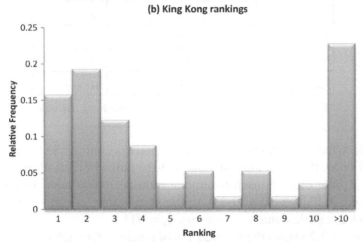

Figure 7.1. Frequency distribution for movie monster rankings.

For example, suppose the movie monster rankings are truly random and the concentration of observations in the top three spots is due to random chance. In this case, the frequency distributions follow a discrete uniform distribution. First shown in Figure 5.2b, the discrete uniform distribution is a common probability distribution where all values have the same probability of occurring. A six-sided die, for example, follows a uniform distribution with possible values ranging from one to six. If it

Ranking	Frequency of rankings (N = 57)		
	Observed for Godzilla	Observed for King Kong	Expected from uniform distribution
1	14	9	5.2
2	9	11	5.2
3	6	7	5.2
4	5	5	5.2
5	5	2	5.2
6	3	3	5.2
7	0	1	5.2
8	2	3	5.2
9	4	1	5.2
10	1	2	5.2
>10	8	13	5.2

Figure 7.2.   Frequency table of Godzilla and Kong rankings.

is indeed the case that both monsters' rankings conform to the uniform distribution, then they are just as likely to be ranked #10 as #1, indicating there is little or no consensus among monster movie fans. To test this hypothesis, I'll set all of the q_k to the same value, and since there are $M = 11$ categories, that means every $q_k = 1/11$.

Figure 7.2 shows the frequencies of rankings for each monster from my sample of data. For Godzilla, these frequencies range from zero to 14. For Kong, they range from one to thirteen. I've also included the frequencies you'd expect if the rankings were purely random, in other words, the result of rolling an 11-sided die 57 times. This expected value is just the number of rolls, 57, multiplied by the uniform probability for each ranking, $q_k = 1/11$.

The test statistic for a goodness-of-fit test measures the distance, or deviation between the observed frequencies and the frequencies you'd expect if your data followed the discrete uniform distribution. For a population with M categories, the formula for this test statistic is

$$\chi^2 = \frac{(O_1 - E_1)^2}{E_1} + \frac{(O_2 - E_2)^2}{E_2} + \cdots + \frac{(O_M - E_M)^2}{E_M}.$$

Here O_k refers to the observed frequencies (column one or two of the table in Figure 7.2) and E_k refers to the expected frequencies (column three).The test statistic for Godzilla's frequencies is

$$\chi^2_{GODZILLA} = \frac{(14 - 5.2)^2}{5.2} + \frac{(9 - 5.2)^2}{5.2} + \cdots + \frac{(8 - 5.2)^2}{5.2} = 31.2.$$

And for King Kong, this test statistic is

$$\chi^2_{KONG} = \frac{(9 - 5.2)^2}{5.2} + \frac{(11 - 5.2)^2}{5.2} + \cdots + \frac{(13 - 5.2)^2}{5.2} = 34.3.$$

A side note here. Because we're comparing the movie monster rankings to the uniform distribution, the expected frequencies for each category are all the same. But they don't have to be. In other words, they can be different from category to category. This χ^2 statistic can be used to test any discrete distribution, as long as the probabilities (and therefore expected frequencies) are all positive and they sum to one. To calculate the expected frequencies for any discrete probability distribution, you'd multiply the category probabilities by the total number of observations.

The statistic χ^2 is called a chi-squared statistic. And if H_0 is true, this statistic approximately follows the following sample distribution.

Probability distribution of the chi-squared test statistic: When all of the expected frequencies are at least five, and when H_0 is true, the chi-squared statistic approximately follows the chi-squared distribution with $M - 1$ degrees of freedom, where M is the number of categories in the population.

Figure 7.3 plots the chi-squared distribution for $M = 11$ categories (ten degrees of freedom). This distribution is decidedly non-normal. Because the test statistic is a sum of squared distances, the value must always be positive. It's also right-skewed, with a long tail. Such is always the case for the chi-squared distribution.

The test statistic χ^2 measures the squared distance between the observed frequencies and those you'd expect if H_0 were true. If this

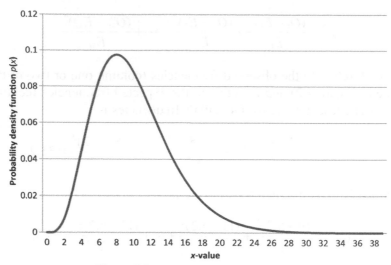

Figure 7.3. The chi-squared distribution.

distance is small, then the evidence is strongly in favor of the null hypothesis. If this distance is large, then the evidence is in favor of H_A. How large is large? That's what the decision threshold tells us. If χ^2 is smaller than the decision threshold, then we accept H_0. Otherwise, we reject this null hypothesis in favor of H_A.

For all hypothesis tests, the decision threshold is determined by the error probability for the test. An *error probability* is the probability you'll make the wrong conclusion, and it's a value you specify when performing the test. There are two types of error probabilities: (1) the probability you'll reject H_0 when H_0 is true, and (2) the probability you'll accept H_0 when H_A is true. The decision threshold comes from the first error probability, what we call the *type I error probability*.

Suppose you'd like the probability you'll wrongly reject H_0 to be less than some small value α. The test is set up so that you reject H_0 if $\chi^2 \geq k_{crit}$, and so this error probability can be written as

$$P\{\text{reject } H_0 \mid H_0 \text{ true}\} = P\{\chi^2 \geq k_{crit} \mid H_0 \text{ true}\} < \alpha.$$

This is a conditional probability *given* the null hypothesis is true, and so the sample distribution for χ^2 tells us this test statistic follows the chi-squared distribution.

Since you know the probability distribution for χ^2, you can use it to find the critical value k_{crit}, and this process is just like the one used for confidence intervals in the last chapter. There's only one difference. For confidence intervals, you use a two-tailed critical value because you're looking for a range of numbers that bound the margin of error for your estimate. For a goodness-of-fit test, you use a one-tailed critical value because you only care about the right tail, the area where the chi-squared distance is too large to be the result of random chance. One-tailed critical values for common α and different degrees of freedom are provided in Appendix C for the chi-squared distribution. For the Godzilla and King Kong rankings, the number of categories is 11. Setting the error probability $\alpha = 0.05$ and the degrees of freedom to 10 for this test gives $k_{crit} = 18.3$.

The rest of the hypothesis test is simple. You just take your test statistic χ^2 and compare it to k_{crit}. If $\chi^2 \leq k_{crit}$, then accept the null hypothesis. If $\chi^2 > k_{crit}$, then reject H_0 in favor of H_A. For the monster rankings, $k_{crit} = 18.31$. $\chi^2_{GODZILLA} = 31.1$ and $\chi^2_{KONG} = 34.2$, both of which are greater than k_{crit}. So in both cases, you'd reject the null hypothesis that the monster rankings follow a uniform distribution. In other words, neither Godzilla's rankings nor King Kong's rankings are consistent with the rolling of an 11-sided die. The differences in frequencies seen in Figure 7.1 are real and not the result of random fluctuation.

For all hypothesis tests, basic or advanced, this process is the same. A test statistic is calculated and compared with the critical value for the relevant sample distribution. However, often you'll see the results reported as a p-value and not some test statistic/decision threshold comparison. A p-value is a tail probability, just like the error probability used to construct the decision threshold. The p-value measures the likelihood of the test statistic being what it is purely by chance. In other words, the p-value expresses the probability, under the null hypothesis, that your test statistic is at least as large as what you observed.

$$p\text{-value} = P\{\chi^2 \geq \chi^2_{observed} \mid H_0 \text{ true}\}.$$

If your test statistic is small, then the p-value will be large. This means that the value of your test statistic has a high probability under the null hypothesis, and this gives weight to H_0. If your test statistic is large, the p-value is small. In this case, the value of your test statistic is unlikely under the null hypothesis, giving more weight to H_A.

Because it's based on a probability and not a hard-to-interpret test statistic value, the p-value is often used in place of a test statistic to perform a hypothesis test. This process is simple. If the p-value is larger than α, your test statistic is reasonably likely under the null hypothesis, and so there is not enough evidence to reject H_0. If the p-value is smaller than α, then the test statistic highly unlikely under the null hypothesis, and so H_0 should be rejected in favor of H_A. Even though the two processes are different, they always end up with the same conclusion. This is because the p-value is calculated from the test statistic and k_{crit} is calculated from α. This relationship between the test statistic, the decision threshold, α, and the p-value are illustrated in Figure 7.4.

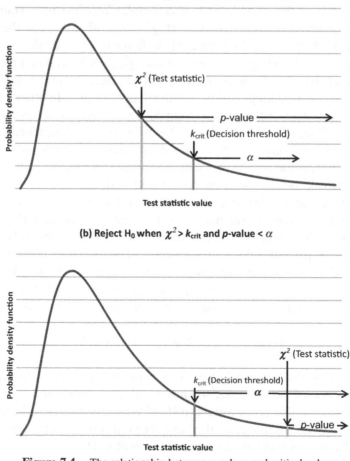

Figure 7.4.   The relationship between p-values and critical values.

Tests for the Mean of a Population

The goodness-of-fit test tells us whether the monster rankings fit a uniform distribution or not, but they don't determine which monster wins the popularity contest. For this, you need a head-to-head comparison, a test that can tell you if the mean ranking of the two monsters is the same or different.

There are several hypothesis tests for the mean of a population, and the one to use depends on your specific hypotheses and whether or not you know the variance of the population beforehand. All of the most common tests assume your data follow a normal distribution, but thanks to the central limit theorem, they work for non-normal populations as well, so long as you have more than about 25 observations in your sample.

Known Variance

The simplest test for the mean works when the variance of your population is known beforehand, so for the moment, let's assume I know this to be σ^2. Suppose I want to test the following claims:

H_0: $\mu = \mu_0$ vs.

H_A: $\mu > \mu_0$.

Like the confidence interval for a mean with known variance, this hypothesis test relies on the z-statistic discussed in the last two chapters. In other words, if $\bar{x}$ is your sample mean, μ is the population mean, σ^2 is the population variance, and N is the number of observations in your sample, then $Z = \sqrt{N}\,(\bar{x} - \mu)/\sigma$ has a standard normal distribution. This gives way to the test statistic for this hypothesis test, which is

$$Z = \sqrt{N}\,\frac{\bar{x} - \mu_0}{\sigma}.$$

Because I'm looking to determine if the mean is equal to or greater than μ_0, the decision threshold will be some value z_{crit}, where if $Z > z_{crit}$, the null hypothesis will be rejected an I'll conclude $\mu > \mu_0$. Just like a goodness-of-fit test, the critical value z_{crit} is determined by the error probability you want to achieve

$$P\{Z > z_{crit} \mid H_0 \text{ true}\} = \alpha.$$

When H_0 is true (as it is in this conditional probability), Z follows a standard normal distribution and so the critical threshold z_{crit} can be calculated as a one-tailed critical value for the standard normal distribution. This is provided in Appendix A.

For example, suppose you have a dataset with sample mean $\bar{x} = 1$, population variance $\sigma^2 = 4$, and number of observations $N = 25$, and you'd like to test the hypothesis H_0: $\mu = 0$ versus H_A: $\mu > 0$. The test statistic is $Z = \sqrt{25}\,(1-0)/2 = 2.5$. If Z is large enough, in other words, larger than the decision threshold z_{crit}, then the null hypothesis will be rejected and you'll conclude $\mu > 0$. Otherwise, you'll stick with the null hypothesis that $\mu = 0$. For $\alpha = 0.05$, the one-tailed critical value $z_{crit} = 1.64$. Because the test statistic $Z = 2.5$ is larger than this decision threshold, the null hypothesis is rejected and I conclude that $\mu > 0$.

The alternate hypothesis H_A: $\mu > \mu_0$ is just one of many possible variations people use in testing the mean. Instead of looking at the claim H_A: $\mu > \mu_0$, you might be interested in the claim H_A: $\mu < \mu_0$. Or, you might not care whether the mean is greater than or less than μ_0, only different, in which case you'd be most interested in the alternate hypothesis H_A: $\mu \neq \mu_0$. Each of these alternate hypotheses uses the exact same z-statistic as the test statistic, but the decision threshold for each is different. For example, to test the hypotheses H_0: $\mu = 0$ versus H_A: $\mu < 0$, you'd look for the Z statistic to be too small to occur by chance; in fact, you'd look for it to be less than zero. So, the decision criterion for this test would be to reject the null hypothesis if $Z < -z_{crit}$, the one-tailed critical value for the standard normal distribution. To test the claim that H_A: $\mu \neq 0$, you'd use the same test statistic, but in this case you'd look for Z to be excessively large or small, in other words, greater than z_{crit} or less than $-z_{crit}$. In this case, you'd use the two-tailed critical value and reject the null hypothesis if your test statistic was outside the range $-z_{crit}$ to z_{crit}. These different variations on a hypothesis test will be illustrated in sections to come.

Unknown Variance: The t-Test

In the real world, if you don't know the population mean, you usually don't know the population variance. This means you must use a hypothesis test for unknown variance. The test statistic in this case is very similar to the previous case where the variance is known; you simply substitute the sample variance for the population variance, giving a

t-statistic instead of a z-statistic. Because a t-statistic has a Student t-distribution under the null hypothesis, you then calculate the critical value using this probability distribution. If $\bar{x}$ is your sample mean, s^2 is the sample variance, and N is the number of observations in your sample, then the test statistic is

$$T = \sqrt{N}\,\frac{\bar{x} - \mu_0}{s}$$

and the decision threshold for this test can be calculated using the inverse t-distribution function.

Suppose you have a dataset with sample mean $\bar{x} = 1$, *sample* variance $s^2 = 4$, and number of observations $N = 25$, and you'd like to test the hypothesis H_0: $\mu = 0$ versus H_A: $\mu > 0$. The test statistic is $T = \sqrt{25}\,(1-0)/2 = 2.5$. If T is large enough, in other words, larger than a decision threshold t_{crit}, then the null hypothesis will be rejected and you'll conclude $\mu > 0$. Otherwise, you'll accept that the mean of your population is zero. The one-tailed critical value for the t-distribution with $\alpha = 0.05$ and $N - 1 = 24$ degrees of freedom is 1.71. Because the test statistic $Z = 2.5$ is larger than this threshold of 1.71, the null hypothesis is rejected and I conclude that $\mu > 0$.

This test for the mean of a population is only set up to compare a single sample to some pre-specified constant value like $\mu_0 = 0$. With a little data manipulation, however, it can be used to compare Godzilla to King Kong. This is because the two monsters' rankings are *paired*. Paired data are exactly what the term implies, observations collected in pairs. For each list, there's a ranking for Godzilla and a ranking for Kong. They are not statistically independent of one another because the ranking of one affects the probability of rankings for the other. For example, if one of the monsters ranks #1 on a list, the other cannot also be ranked #1 (unless ties are allowed, which never happened in the lists I sampled).

When data are paired in this way, we can compare them by taking the difference between the pairs and performing a hypothesis test for the mean on the result. When the population variance is unknown, this type of test is called a *paired t-test*. This is a little different from a generic *t-test* where the two samples are independent of one another. Both tests rely on a t-statistic and both use Student's t-distribution to calculate the critical value, but a little extra work is needed when calculating the sample variance for the generic t-test. For the details on the *t*-test for independent samples, I refer you to one of the texts listed at the end of this book.

To perform a paired *t*-test, I'll redefine my data values to be d_{G-KK}, the difference between Godzilla's ranking and King Kong's ranking on the same list. If the difference is negative, Godzilla ranks higher on the list than Kong. If it's positive, the reverse is true. I'll construct a test for the mean difference between the monster's rankings, in other words, for the following two hypotheses:

H_0: $\mu_{G-KK} = 0$ vs.

H_A: $\mu_{G-KK} \neq 0$.

If the null hypothesis, H_0, is true, then there is no real difference in the rankings between the two monsters, and the battle is a tie. If H_A is true, then one of the monsters ranks, on average, higher than the other and we have a winner.

Before I run the hypothesis test, there's one important glitch in the data to deal with. Thirty-six percent of the lists in my sample ranked one of the monsters in the top ten and not the other. This means one of the monsters ranked lower than ten, but there's no way of knowing how much lower. When observation values are hidden in this way, they are *censored*. There are different ways of dealing with censored data. For example, I could ignore all of the lists where only one of the monsters appears. But this feels like cheating. The fact that one monster didn't appear in the top ten is useful information and should, if possible, be taken into account. So I'll use another common method for handling censored data, I'll set the value of those censored rankings as close as possible to the cut-off value of ten. In other words, whenever Godzilla or King Kong doesn't appear on somebody's top ten list, I'll set his ranking to 11.

Figure 7.5 plots the frequency distribution for the difference between Godzilla's rank and King Kong's rank. These differences are not continuous, and they're definitely not bell-shaped. Because there were no ties in the rankings, there are positive differences and negative differences, but no zero values. The differences are cut off at −10 and 10 because of the way in which I substituted 11 for every censored value. Fortunately, I have well over 25 values in my dataset and I have the central limit theorem. This life-saving property suggests that no matter how beastly the original data are, the *sample mean* tends to look

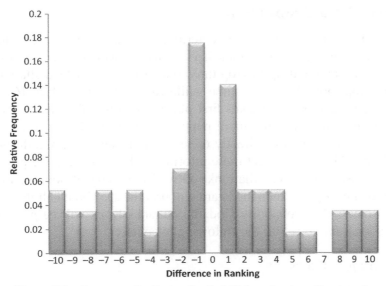

Figure 7.5.   Frequency distribution for Godzilla's rank minus Kong's rank.

approximately normal for large sample sizes. So I can feel comfortable about using the test despite the decidedly non-normal looking data shown in Figure 7.5.

The sample mean of the differences is $\bar{x} = -0.71$. This means, on average, Godzilla's ranking tends to be just slightly lower than Kong's, giving the lizard the edge in the competition. Is this enough of a difference to claim statistical significance? A paired t-test will tell us. The sample standard deviation is $s = 5.28$. The number of observations in the sample is $N = 57$, so the test statistic is

$$T = \sqrt{N}\,\frac{\bar{x} - \mu_0}{s} = \sqrt{57}\,\frac{-0.71 - 0}{5.28} = -1.0.$$

The alternate hypothesis states $\mu_{\text{G-KK}} \neq 0$, and so a two-tailed critical value applies here. From Appendix B, the two-tailed critical value for the t-distribution with $\alpha = 0.05$, and $N - 1 = 56$ degrees of freedom is $t_{\text{crit}} = 2.0$. The test statistic $T = -1.0$ is between $-t_{\text{crit}}$ and t_{crit}, not large or small enough to reject the null hypothesis, and so I accept the claim that the mean difference is zero. In other words, I'm forced to stick with the hypothesis that the popularity of the two monsters is the same.

Tests for a Proportion

A tie. How utterly boring. I'm determined to find a winner, regardless of who it might be. As I look back at the frequency distributions in Figure 7.1 and add up the top three bins, I notice that Godzilla ranks in the top spot 25% of the time while Kong takes first place 16% of the time. That Godzilla takes first place more often than Kong might be an indication he's the more popular monster. In the last section, I compared the rankings directly to see if the means were significantly different. Maybe I can get more information by looking at the proportion of lists where Godzilla ranks above King Kong. If both monsters are equally popular, then the lizard will beat the ape about half the time. This means the probability Godzilla outranks Kong will be $p_G = 0.5$. This leads me to set up the following hypotheses:

H_0: $p_G = 0.5$ vs.

H_A: $p_G > 0.5$.

Just like the confidence interval for a proportion introduced in the last chapter, this test relies on the binomial approximation to the normal distribution. In other words, as long as $N\hat{p} \geq 5$ and $N(1 - \hat{p}) \geq 5$,

$$Z = \frac{N\hat{p} - p_G}{\sqrt{N\hat{p}(1 - \hat{p})}}$$

approximately follows the standard normal distribution. With $p_G = 0.5$ and $N = 57$, both of these conditions hold, and so the z-statistic can be used as a test statistic for this test. Since the alternate hypothesis is $p_G > 0.5$, a one-tailed critical value applies here, and the decision threshold can be retrieved from the table in Appendix A.

In my sample, Godzilla beats King Kong 56% of the time; this means $\hat{p}_G = 0.56$. With N = 57 samples, this gives a test statistic of

$$Z = \frac{\hat{p}_G - p_G}{\sqrt{\dfrac{p_G(1 - p_G)}{N}}} = \frac{(0.56) - 0.5}{\sqrt{\dfrac{(0.5)(0.5)}{57}}} = 0.93.$$

From Appendix A, the one-tailed critical value for the standard normal distribution with $\alpha = 0.05$ is $z_{crit} = 1.65$. Because the test statistic Z is less that z_{crit}, I must, once again, accept the null hypothesis that the monsters are equal.

AND THE WINNER IS . . .

Sometimes a study doesn't support our preconceived notions or prove our point. This is one of those times. I wanted King Kong to win. I thought Godzilla was going to win. In the end, neither monster came out on top. After analyzing the monster ranking with three of the hypothesis tests summarized in Figure 7.6, I'm forced to draw the unsatisfying conclusion that both monsters are equally popular among movie fans. It's possible, with a larger sample size, that the variation of the test statistics could be decreased enough for one of these creatures to win. It's

Type of test	H0	HA	Test statistic	Reject H0 when:	Where to find the critical value
Mean, known variance	$\mu = \mu_0$	$\mu > \mu_0$ $\mu < \mu_0$ $\mu \neq \mu_0$	$Z = \sqrt{N}\dfrac{\bar{x} - \mu_0}{\sigma}$	$Z > z_{crit}$ $Z < -z_{crit}$ $Z > z_{crit}$ or $Z < -z_{crit}$	Appendix A
Mean, unknown variance	$\mu = \mu_0$	$\mu > \mu_0$ $\mu < \mu_0$ $\mu \neq \mu_0$	$T = \sqrt{N}\dfrac{\bar{x} - \mu_0}{S}$	$T > t_{crit}$ $T < -t_{crit}$ $T > t_{crit}$ or $T < -t_{crit}$	Appendix B
Proportion	$p = p_0$	$p > p_0$ $p < p_0$ $p \neq p_0$	$Z = \dfrac{N\hat{p} - p}{\sqrt{N\hat{p}(1 - \hat{p})}}$	$Z > z_{crit}$ $Z < -z_{crit}$ $Z > z_{crit}$ or $Z < -z_{crit}$	Appendix A
Chi-squared (goodness-of-fit)	$p_1 = q_1,$ $p_2 = q_2,$ etc.	At least one $p_k \neq q_k$	$\chi^2 = \dfrac{(O_1 - E_1)^2}{E_1} + \cdots$ $\cdots + \dfrac{(O_M - E_M)^2}{E_M}$	$\chi^2 > x_{crit}$	Appendix C

Figure 7.6. Common hypothesis tests for common situations.

also possible that there is no consensus among movie fans and so no amount of top ten lists would ever result in a statistically significant difference. If you're so inclined, I urge you to gather twice as many top ten lists and repeat this analysis to see if the lizard or the ape wins. If you do declare a winner, let the world know. Until then, the battle of the movie monsters rages on.

BIBLIOGRAPHY

BBC NEWS. "King Kong Tops Movie Monster Polls." Accessed July 9, 2012, http://news.bbc.co.uk/2/hi/entertainment/3596551.stm.

IMDB.COM. "Plot Summary for Godzilla." Accessed October 12, 2012, http://www.imdb.com/title/tt0047034/plotsummary.

IMDB.COM. "Plot Summary for King Kong." Accessed October 12, 2012, http://www.imdb.com/title/tt0024216/plotsummary.

REELZCHANNEL TV. "Hollywood's Top Ten Movie Monsters." Accessed April 25, 2012, http://www.reelz.com/trailer-clips/56346/hollywoods-top-ten-monster-movies/.

CHAPTER 8

Lab Rats and Roommates: Analysis of Variance and How to Channel Your Inner Mad Scientist

I have a confession to make. I'm a bit of a mad scientist.

I was the kid with the 1001 Electronics Kit and the Early Learners Chemistry Set and the Little Critters Biology Lab. Only I never did the experiments in the books that came with those kits. No, I had my own set of experiments to run, my own list of questions to answer. Like whether or not my little brother could feel an electrical current passed through the doorknob on the door to his bedroom (he could). I was the girl in high school who helped the chemistry teacher prepare mixtures for classroom demonstrations. But it wasn't mixing solutions and weighing chemicals that interested me. I wanted answers to questions that didn't appear in my textbook. Like whether or not silver nitrate really does turn your best friend's feet black (it does). I was the chemistry student in college whose labs never worked out because I was too busy testing other hypotheses. Like the TA's claim that a six-inch Bunsen Burner can't create a flame high enough to reach the ceiling (actually, given the right motivation, it can).

The Art of Data Analysis: How to Answer Almost Any Question Using Basic Statistics, First Edition. Kristin H. Jarman.
© 2013 John Wiley & Sons, Inc. Published 2013 by John Wiley & Sons, Inc.

Being a mad scientist has many benefits. You never have to worry about a bad hair day. The badder, the better. You can test any theory without worrying about those annoying details that bother other people, details like *ethics* and *other people's feelings*. And if you're lucky enough to live with family members or roommates, you have an endless supply of test subjects at your disposal. All you need is a hypothesis, a controlled experiment, and a data analysis plan and all your nagging questions can be answered.

Nowhere is mad science more evident, or more socially acceptable, than a middle school science fair. Pliable young minds grow mold in their refrigerator, subject their siblings to any manner of tests, and feed their pets strange foods just to answer some question that popped into their heads. And creativity is encouraged. Projects are supposed to be scored on the merit of the work and not on the student's hypothesis. So it shouldn't matter that Dillon contaminated the entire contents of his mother's pantry with ants. As long as he followed the scientific method and came to a valid conclusion, his project is just as good as any other.

But is it, really?

The judges are scientists, and as such, have been trained to be objective. But judges are also people, complete with their own imperfections and personal biases. Does Professor Bennett let her own fear of insects cloud her view of Dillon's brilliant hypothesis? Does Dr. Lincoln look beyond the ant infestation and into the teenager's insightful conclusions? In this chapter, the scientists become the lab rats as I ask the question, "Are science fair judges really fair?"

ANALYSIS OF VARIANCE: THE SCIENTIST'S TOOLKIT

The scientific method is the foundation of modern research. It's how we prove a theory. It's how we demonstrate cause and effect. It's how we discover, innovate, and invent.

There are five basic steps to the scientific method:

1. Ask a question.
2. Conduct background research.
3. Come up with a hypothesis.

4. Test the hypothesis with data.

5. Revise and retest the hypothesis until a conclusion can be made.

The majority of this book is devoted to step four, collecting and analyzing data to prove or disprove your hypothesis. In this chapter, however, I'd like to focus a little attention on the entire process. Suppose I've just developed a new, genetically engineered chili pepper. This pepper, the Nine-Levels-of-Hell pepper, is so hot it could, quite literally, set your tongue on fire. As a mad scientist, I'm not happy with the notion that it's the hottest chili ever produced. I need to prove it. So, I set down the path of the scientific method, trying to answer the question, "Is the Nine-Levels-of-Hell pepper the hottest chili pepper known to man?"

I start with some background research. As it turns out, there are two methods people use to determine the heat of a chili pepper (Bosland 2010). The first method involves careful measurements using a sophisticated laboratory instrument called a high-performance liquid chromatograph, or an HPLC. I don't keep one of those in my kitchen, so I immediately turn to the second. This method involves human taste testers and something called the Scoville scale. Dried chili peppers are ground up and dissolved in sugar water like some sort of tongue-numbing cocktail. The cocktail is diluted little by little until the tasters claim they can't taste the heat anymore. The amount of dilution needed to completely wash out the burn is the heat of the chili pepper in Scoville units. A typical green bell pepper is 0 units. A jalapeno pepper is about 4,000 units. A habanero pepper is around 250,000 Scoville units. And the world record holder, the Indian ghost chili pepper, is around 1,000,000 Scoville units (The Scoville Heat Measurement Chart 2012).

All of these peppers are wimps compared to my Nine-Levels-of-Hell chili pepper. At least, that's my hypothesis. And to test it, I can follow the experimental design and planning process outlined in Chapter 2.

The effect in this experiment will be the heat of different chili peppers as measured by the Scoville scale. Since there's a standard procedure for mixing and tasting the Scoville cocktails, all the variables related to preparing the drinks have already been systematically set for me. Even so, there are a number of variables that need to be decided upon. First, I need to pick test subjects, people who are willing to taste each of the chili pepper cocktails and give me feedback. A group of people who start to sweat while eating pepperoni will have a different threshold for heat than people who eat jalapenos by the fistful. This variable, spice tolerance, could definitely have an impact on my results. To keep this uncontrollable factor from becoming a confounding factor, it's important for me to solicit test subjects at random, and gather information from each volunteer about his or her spice tolerance.

There are several controllable factors that need to be set. The types of chili peppers I choose to include in the experiment, who tastes which chili peppers, the order in which the different peppers will be tested, all these are controllable factors. As for the chili peppers to include, I'll pick two of the most infamous competitors out there, the ghost chili pepper and my personal nemesis, the habanero. I'll test these two along with my Nine-Levels-of-Hell pepper and compare all three.

There are two ways to decide who tastes which chili peppers. One would be to have all my test subjects taste all three chili peppers. This would produce a set of paired samples, where the feedback of each test subject is paired across all three groups. There are two problems with this approach. First, except for the paired *t*-test from last chapter, most basic statistical analysis methods assume independent groups and not paired samples. Having feedback from the same individuals in each group makes the samples dependent on one another, and unless I'm willing to apply a not-so-basic statistical analysis technique to deal with this dependency, it's better to avoid this situation altogether. Second, experience has shown that most people only agree to be my test subject once. After tasting their first Scoville cocktail, it's quite likely my test subjects will leave, never to speak to me again. So, rather than having a small number of test subjects, each tasting every chili pepper, I'll gather more test subjects and have each person taste only one. This will give me three independent groups to work with. But in order to keep bias from creeping into my data, I'll assign each test subject to one of the chili pepper groups at random.

All that remains is to formulate a plan for analyzing the data. There are three chili peppers in the experiment, and so there will be three groups to compare to one another. Because there are more than two groups, a simple hypothesis test isn't enough. *Analysis of variance* is the tool for the job. Analysis of variance, or ANOVA, is just a fancy term for a hypothesis test that compares the mean of multiple groups. Like the hypothesis tests described in the last chapter, ANOVA begins with a null hypothesis and ends with the comparison of a test statistic (or *p*-value) to a decision threshold (or error probability α). However, it's specifically designed to compare the means of more than two groups at once.

The *t*-test and ANOVA are both hypothesis tests for equality of the mean of different groups. If there are two groups, then a *t*-test should be used. If there are three or more, then ANOVA is the tool of choice.

Suppose I have samples from three different populations, and these three populations have means μ_1, μ_2, and μ_3. To test for sameness of all three, the null hypothesis looks like

$$H_0: \mu_1 = \mu_2 = \mu_3.$$

There are plenty of tests for comparing the means of two populations. And if you were willing to compare all possible pairwise combinations, 1 versus 2, 1 versus 3, and 2 versus 3, you could use a *t*-test to compare these three groups to one another. This is called a *multiple comparison procedure*. Multiple comparison procedures can be helpful in determining which groups are different from others, and it's a great tool for analyzing many groups in detail. But the process is cumbersome and the error probabilities can be tricky; the more groups you have, the worse it gets. For example, with four groups, there are six combinations of two groups to take into account. With five groups, there are ten. With ten groups, there are 45.

ANOVA offers a simple and powerful alternative to this all pairwise comparison approach. ANOVA doesn't look at group means two-by-two, rather it tests for sameness of all group means at once. It does this indirectly, by looking at variation, relying on a common hypothesis test for comparing the variance of two populations. This test is called the F-test, and I'll introduce it before moving on to the ANOVA procedure.

Suppose you have two groups and you'd like to know if the variance of those two groups is the same. You could set up the following hypotheses:

$$H_0: \sigma_1^2 = \sigma_2^2 \text{ vs.}$$
$$H_0: \sigma_1^2 > \sigma_2^2.$$

Like all the hypothesis tests in the previous chapter, the test statistic relies on a key result about a sample distribution. Here's the result:

Probability distribution for the F-statistic: For two groups with the same population variance, sample variance s_1^2 and s_2^2, and number of samples N_1 and N_2, the *F-statistic*, $F = s_1^2/s_2^2$, follows

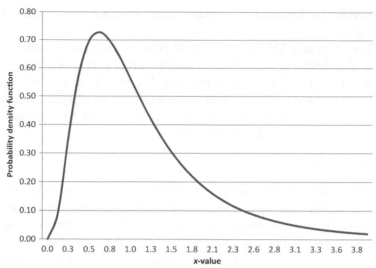

Figure 8.1.   The F-distribution.

the *F-distribution* with $N_1 - 1$ degrees of freedom in the numerator and $N_2 - 1$ degrees of freedom in the denominator.

The F-distribution is plotted in Figure 8.1 for $N_1 = N_2 = 10$. Like the chi-squared distribution from the last chapter, this distribution is decidedly non-normal. Because the variance is always positive, the values for the F-distribution are always positive as well. It's right-skewed. The mean is always one, and the variance depends on the degrees of freedom in the numerator and the denominator.

The F-statistic is the test statistic for the F-test, and so the one-tailed decision threshold F_{crit} for this test can be calculated using critical values from the F-distribution. If $F \leq F_{crit}$, then you accept the null hypothesis that the group variances are the same. If $F > F_{crit}$, then you reject the null hypothesis and conclude that they are different.

I haven't provided a table of critical values for the F-distribution because as you'll see later in this chapter, the ANOVA procedures available in most data analysis software packages provide this for you. However, if you're so inclined, critical values for the F-distribution can be calculated in Excel using the FINV function.

To compare the variance of two groups with an F-test, you label the group with the larger sample variance as group one, calculate the

F-statistic, and compare it to the decision threshold. For example, suppose I have two groups, each with $N_1 = N_2 = 25$ samples in them. The sample variances are $s_1^2 = 19.4$ and $s_2^2 = 16.2$. (Note that I set it up so group one has the larger variance.) The F-statistic is F = 19.4/16.2 = 1.19. From Excel, the $\alpha = 0.05$ critical value for the F-distribution with $N_1 - 1 = 24$ degrees of freedom in the numerator and $N_2 - 1 = 24$ degrees of freedom in the denominator is $F_{crit} = 1.98$. Because $F < F_{crit}$, I accept the null hypothesis that the variances of the two groups are the same.

So, what does an F-test have to do with ANOVA? How can a test for sameness of variance be used to compare three or more means? When you compare two or more groups, there are different sources of variation in the data. First, there's the variation of observations within each group. This type of variation, called the within-group variation, is typically measured with the sample standard deviation. There's also variation between the different groups. This between-group variation comes not from the spread of observations within a data cloud, but from differences between the central location of the different clouds. The between-group variation is typically measured using the standard deviation of the group averages. These different sources of variation are illustrated for three groups in Figure 8.2. Note the brackets showing within- and between-group variation are meant to illustrate the concept of variation and not the standard deviation, because standard deviation measures the average deviation around the sample mean and not the total spread of observations as suggested in the figure.

The ANOVA procedure tests for equality of group means by performing an F-test that compares between-group variance (the numerator) to the typical within-group variance (the denominator). If the between-group variance is significantly larger than the typical within-group variance, the group means vary more than would be expected by random chance alone. In this case, the null hypothesis that the group means are equal is rejected. Otherwise, the null hypothesis is accepted.

The calculations for the ANOVA test are detailed and a little messy, so I'll leave them to the references listed at the end of this book. Fortunately, you don't need to know all the detailed calculations to run ANOVA. This procedure is available in most basic spreadsheet programs such as Microsoft Excel. All you need is your data, organized into groups, a desired significance level α, and a basic understanding of the assumptions that go into such an analysis.

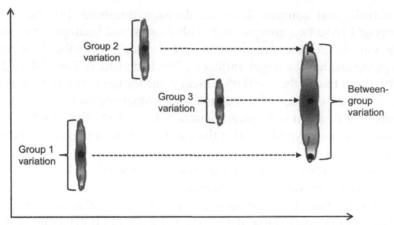

Figure 8.2. Within-group variation versus between-group variation.

ANOVA assumptions:

1. The samples approximately follow a normal distribution.
2. None of the sample standard deviations is dramatically different from the others.
3. The samples are independent of one another, in other words, not collected in groups.

ANOVA results are typically displayed in a table format. For example, suppose I run the chili pepper comparison experiment and get three groups of Scoville measurements, one for each pepper. A typical ANOVA procedure performed on these data might look like the results presented in Figure 8.3. The first table provides a routine summary of each group, including the number of observations, the sample mean, and sample variance. The second table summarizes the results of the F-test. This table reports intermediate values that go into the calculation of the F-statistic, things like the total sum of squared deviation, degrees of freedom, and average squared deviation (the total sum of squared divided by the degrees of freedom). The most important values are reported in the last three columns. The F-statistic is reported in the fourth column. This is the ratio of the between-group to within-group variances. The *p*-value of this F-statistic is reported in the fifth column. The decision threshold for the test is listed in the last column.

Data summary			
Group	Number of observations	Average Scoville units (x100,000)	Variance
Ghost chili	24	91.25	301.76
Habanero	24	56.29	386.82
Nine-Levels-of-Hell	24	114.88	450.64

ANOVA						
Source of Variation	Sum of squared deviation (SS)	Degrees of freedom (df)	Mean squared deviation (MS)	F-statistic	P-value	F crit
Between Groups	34087.19	2	17043.6	44.88	3.27E-13	3.1
Within Groups	26202.08	69	379.74			
Total	60289.28	71				

Figure 8.3.   Sample output of ANOVA for the chili pepper comparison.

For this example, the F-statistic is F = 44.9, and the critical value for the test is F_{crit} = 3.13. Since F > F_{crit}, then the null hypothesis that all three chili peppers are equal is rejected. The second to last column, the p-value, reports just how unlikely this value F is under the null hypothesis. Specifically, if the group means are the same, the probability the F-statistic would have been bigger than 44.9 is the p-value, the exceedingly small probability of 3×10^{-13}. In other words, the probability F ≥ 44.9 purely by chance is so small, the null hypothesis should be rejected.

The three chili peppers aren't the same. The ANOVA results tell us at least one of them is different from the others. But which one is it? While it's possible to look at the different group means and speculate, ANOVA doesn't provide a definitive answer to this question. That's what a multiple comparison procedure does. By comparing groups two-by-two, this procedure allows you to rank the groups from largest to smallest, identifying which groups have the same means and which are significantly different. But this is a subject for another day, and so the details of the multiple comparison procedure are left to the texts listed at the end of this book.

LET THE JUDGES BE JUDGED

There are science fairs and there are Science Fairs. The Intel International Science and Engineering Fair, for example, is a Science Fair. In this competition, high schoolers cure cancer in order to win up to $100,000 in cash and scholarships. Science Fairs showcase the world's future Nobel prize winners. Science Fairs are poster children for the responsible and ethical practice of the scientific method.

I have no interest in Science Fairs.

I'm more interested in a science fair, a local middle school contest where kids practice mad science, the kind that makes them feed energy drinks to plants just to see if BuzzRush is good for you, the kind that gets them to listen to their music player for 85 hours straight just to see how long batteries last. That's the science fair I'm interested in.

After a little searching, I found a local middle school willing to give me scores from their 2012 science fair (and you know who you are, thanks for the data). Here's how the judging went. There were 67 science projects on display, and 27 test subjects, er, judges to score them. Each judge was randomly assigned between seven and nine projects, so that each project received scores from at least three judges. The projects were scored in four categories, each category on a scale from one to five, so the lowest possible score was zero, and the highest possible score was twenty.

All the judges were practicing scientists, with areas of expertise overwhelmingly focused on physics, chemistry, and biology. The student projects covered a much broader range of topics, including behavioral science, food science, plant science, materials, and so on. If the judges were biased for or against certain categories of projects, that bias should be reflected in the final scores. Therefore, the experimental plan was this: to group the projects by category and compare the scores across different groups, looking for differences that suggest the judges are less than fair.

This strategy is not without its problems. All project categories are not created equal. Some tend to be more scientifically demanding than others. Engineering, for example, is a difficult subject. And while it might appeal to the young Albert Einstein out there, the kid who's been building a mini nuclear reactor in his own back yard, it probably doesn't appeal to the kid who pays his lab partner to do the experiments while he sits in the back of the room doodling. If this is indeed the case, then the scores in the engineering category might tend to be higher than others, not because of any bias on the part of judges, but because the more scientifically-skilled students tend to prefer it.

This variable, the relative difficulty of different categories, has the potential to become a confounding factor. In other words, if I ignore it, this factor could make the conclusions of this experiment ambiguous. As discussed in Chapter 2, there are several ways to eliminate the impact of confounding factors. One is randomization and another is blocking. Randomization could be used to remove this potential problem, but the school frowned upon letting me choose project topics for the kids (something about encouraging the kid's curiosity and creativity). So, I had to settle for blocking.

Blocking is the process of organizing the data into blocks that are homogeneous with respect to a potentially confounding factor. The school was willing to give me the classroom grades for each project, these grades being assigned by the teacher. The teacher scored the projects on the scientific method as learned in class, providing an independent assessment of each project. In other words, these grades set the standard for what makes a good and bad project. In this study, I included only those projects that received an A or B, in other words, only the best projects.

I categorized the projects along the lines of the Intel International Science Fair guidelines with the addition of one: food science. This category includes a variety of studies comparing the taste or texture of differently processed foods. The frequency distribution of the project categories is provided in Figure 8.4. The two most common categories are physics and food science. Each of these categories contained 11 projects. The least popular category was health science, with only a single entry.

All projects were scored on a scale from 0 to 20, and Figure 8.5 shows the average scores for each category, along with the total variation as calculated by the range (see Chapter 2 for details). The average, or sample mean, is plotted as a diamond, and the range is illustrated by the vertical line. The data clouds are plotted simply to illustrate the total within-group variation. To the right, the between-group variation is plotted. The grand mean, the average of all the group means, is plotted as a diamond and the range of group means is shown with the vertical line.

According to the figure, the sample mean of most groups falls between 14 and 16. The within-group variation, the span of the

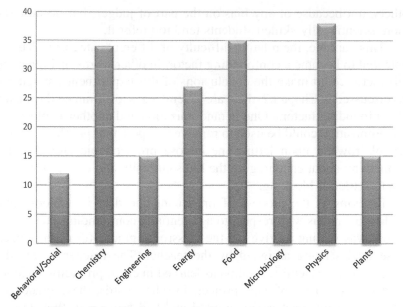

Figure 8.4. The frequency distribution of science fair projects.

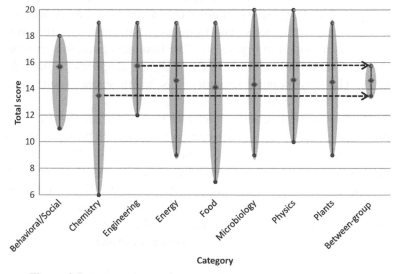

Figure 8.5. The sample mean and total variation of science fair scores.

individual data clouds, easily covers the span of group means, from the lowest scoring chemistry category, to the highest scoring engineering category. In other words, it looks like all the group averages are consistent with one another within the observed variation. However, this figure is merely a visual aid. It doesn't provide for any rigorous analysis nor does it use the recommended measure of variation, the variance. Let's see what ANOVA says.

The ANOVA procedure assumes each of the groups follows a normal distribution, but it's pretty forgiving and so many researchers ignore the assumption and use this procedure anyway. But it's important to note that in such cases, the probability of making a wrong conclusion might be larger than the $\alpha = 0.05$ significance level specified by the test. It might be off by just a few percent, or it might be off by a factor of ten. It all depends on how non-normal your data are. As a general rule, when the assumptions of ANOVA (or any hypothesis test) are violated, it's okay to run the test anyway, but it's a good idea to check your conclusions by looking at a graph such as the one provided in Figure 8.5.

Figure 8.6 shows the results of the ANOVA procedure performed on the science fair data. The last three columns contain the most important results. The F-statistic is 1.50 and the critical value is $F_{crit} = 2.06$. The p-value is 0.17, over three times larger than the significance level $\alpha = 0.05$. Since the p-value is so large and $F < F_{crit}$, I must accept the null hypothesis that the mean scores are the same across categories. These results are consistent with my suspicions from Figure 8.5. It looks like science fair judges really are fair, after all.

UNLEASHING YOUR INNER MAD SCIENTIST

These results should be a comfort to all those mad scientists in training out there. They're definitely a comfort to me. It apparently doesn't

Source of Variation	SS	df	MS	F	P-value	F crit
Between Groups	83.07834	7	11.86833	1.502161	0.16894	2.059914
Within Groups	1445.854	183	7.900839			
Total	1528.932	190				

Figure 8.6.   ANOVA analysis of science fair scores.

matter what type of science you prefer, whether it's physics, food chemistry, or psychology. Nor does it seem to matter if your motives are misunderstood and your creativity underappreciated. Your project will be judged on its scientific merit. And even if you never win thousands of dollars or make it to an internationally recognized competition, you still have the satisfaction of knowing that with a good experimental design and the right statistical tools, channeling your inner mad scientist is as easy as finding a few willing family members and friends.

BIBLIOGRAPHY

BOSLAND, PAUL W. and STEPHANIE J. WALKER. 2010. *Measuring Chili Pepper Heat.* Las Cruces: New Mexico State University Press.

THE SCOVILLE HEAT MEASUREMENT CHART. Accessed August, 7, 2012, http://www.wiw.org/~corey/chile/scoville.html.

SOCIETY FOR SCIENCE AND THE PUBLIC. "INTEL INTERNATIONAL SCIENCE AND ENGINEERING FAIR." Accessed August 6, 2012, http://www.societyforscience.org/page.aspx?pid=270.

CHAPTER 9

When the Zombie Flu Went Viral: Regressing the Myth Out of Urban Myths

One of the great things about the Internet is its ability to spread news at the speed of light. Earthquakes, tsunamis, and hurricanes are watched while they happen. Politicians' criminal activities are posted the moment the money changes hands. Celebrity divorces are reported in the tabloids, even, it seems, before the stars are aware they're unhappy. This rabid information flow doesn't stop with the truth, either. It also applies to tales of a more dubious origin.

Snopes.com is a website devoted to tracking and debunking urban myths. Here you'll find strange but true stories, like the molasses flood of 1919 that killed dozens of Bostonians. You'll learn about current scams, like the famous Nigerian Bank scam, the one where you get an email from somebody claiming to be a rich foreigner who desperately needs to transfer his money into your bank account so his government doesn't take it all. You'll also see completely false reports, urban legends, like the one about the organ-stealing con men who take your kidneys and sell them on the black market. As far as questionable reports go, if you've heard of it, this website probably has it.

The Art of Data Analysis: How to Answer Almost Any Question Using Basic Statistics, First Edition. Kristin H. Jarman.

The zombie flu is just such a report. According to the website, this strange virus first appeared in April 2005, when a fake but official-looking BBC news article announced a new parasite had appeared in Cambodia. Carried by mosquitoes, this deadly parasite reportedly killed its victim, and then restarted the heart for up to two hours after death, causing the unfortunate person to behave like a zombie on the rampage. It was cleverly crafted, including everything from statistics on the mortality rate to official reactions from both Cambodia and the United States.

The report sat on the Internet, infecting the imaginations of a few diehard zombie fans, but it wasn't until four years later when the disease went viral. On April Fool's Day 2009, the report was revived, reworked, and recirculated. Now a mutation of the H1N1 virus, the zombie flu was reportedly spreading through London, with suspected cases in other European cities. The rumor spread like a pandemic. News of the strange disease was sent out on Twitter. Bloggers began writing about the virus and its role in the upcoming Zombie Apocalypse. News-groups filled up with questions from concerned people asking if the zombie flu was real and how they could prepare for it.

According to the website techcrunch.com, the rampant spread of zombie swine flu was sparked by the Twitter community. Labeled as an authentic BBC news story and allowed to be retweeted under the same heading, Twitter followers around the world were among the first to hear about the disease. Was this the trigger, the event that sparked the rumor pandemic? In this chapter, I'll use basic regression to track reports of this grisly condition, showing how a nugget of truth can explode into an urban legend, answering the question "When did the zombie flu go viral?"

LINEAR REGRESSION AND OTHER ZOMBIE TRACKING TOOLS

When you're tracking trends, regression is the tool of choice. *Regression* is a procedure for describing one variable, y, as a function of another variable, x. The variables can be anything as long as there's good reason to assume y depends on x. For example, the total number of human brains a zombie has eaten over time might be turned into a regression function, with the number of brains being the y-variable and time being the x-variable. The number of brains eaten depends on how long the zombie has been on the rampage and so y is a function of x, where $y = f(x)$.

The function relating y to x can also be almost anything—a line, a quadratic function, even a logarithmic function. Depending on the form of this function, whether linear or nonlinear, the regression process ranges from simple to highly sophisticated. The most basic regression procedure, *linear regression*, assumes x and y are related to one another with a line, $y = mx + b$. This is the easiest type of regression to perform, and the one I'll illustrate throughout this chapter.

Basic linear regression takes a set of paired observations, call them x and y, and fits the line $y = mx + b$. In other words, it uses the data to estimate the slope of the line, m, and the y-intercept, b. Because samples and not entire populations are generally used to get these estimates, there's uncertainty associated with them. So, an important part of the regression process is evaluating the uncertainty in these estimates and determining what's called the *prediction error*, the variation inherent in the y-value as predicted by the regression line. Linear regression procedures available in most basic data analysis packages provide not only estimates of m and b, but also a number of useful diagnostic statistics as well. And as you'll see throughout the rest of this chapter, these diagnostics often turn out to be just as important as the estimated parameters.

INTERNET POSTS AND OTHER SYMPTOMS OF DISEASE

When a story, blog, or video goes viral on the Internet, it becomes extremely popular in a very short period of time. Popularity can be measured in several ways, such as the number of hits a website receives, how fast a particular item is being shared, or how often the topic is being discussed. The zombie flu has been around since 2005, and it was already a regular topic of discussion long before zombies became the rage in 2009. If the fake BBC article was the trigger that made the zombie flu go viral, then there should be a dramatic difference between the amount of related posts before and after April Fool's Day 2009.

To look at this, I decided to track Internet activity on the zombie flu during the first half of 2009. Google allows users to search within a range of posting dates, and so a search on the term "zombie flu outbreak" for the first half of 2009 would provide me with a good sample of postings on the topic during the months in question. My plan was to grab the dates of each posting and look at the number over time. Before going viral, this number should increase at a nice, steady pace. After going viral, the rate of new postings should really take off.

WHEN THE ZOMBIE FLU WENT VIRAL

Did you know Google watches you when you're web searching? Sure, I'd heard this was true, but I didn't really believe it until I starting collecting search results on the zombie flu. See, I bought a web ripper. It's a piece of software that allows you to automatically extract selected bits of information from any website. After only one tutorial and a couple hours, I had a basic working knowledge of the software. I must admit, it gave me a sense of power, knowing I could extract all sorts of information from any website. And I didn't even need to be sitting at my computer to do it. I could set up the program to automatically scroll through pages of search results, grabbing any information I wanted. All it took was a little money and a little willingness to learn some new software.

After searching "zombie flu outbreak" postings between January 1 and May 1, 2009, I set up my web ripper to automatically scan through hundreds of pages of search results, pulling out the link, the summary information, and the posting date of every single hit. Naturally, the software did this much faster than I could have myself. It would have taken me days, but the program could do it in just a few minutes.

Apparently my web ripper was so fast it sent up some red flags in the Google software. Out of nowhere, my program halted as the search engine diverted me to a new web page, announcing it was concerned about the high level of activity I was generating. To continue, the web page said, I'd need to type in a CAPTCHA, a nonsense word written in squiggly letters meant to prevent people like me from releasing viruses and scams on the Internet. I followed the instructions, typing in the phrase, but it only asked me to enter another one. I typed the next nonsense word and it asked for another. It went on like this for a few minutes until I finally gave up. I turned off the software and went for a cup of coffee.

I came back a little while later to continue my web ripping and was delighted to find it working once again. I started the program, sat back, and relaxed. But after just a few minutes, Google stopped me a second time. Like before, I was redirected, lectured, and asked to play the enter-the-CAPTCHA-phrase game once more. This time, I didn't bother. I exited the software and went for another cup of coffee. It went on like this for the better part of a day, me grabbing a couple dozen pages of results and Google eventually stopping me. I'm sure the company was only doing this to protect itself from computer viruses and such, but after four tries and just as many lattes, the paranoia center of my brain launched into overdrive. If it's so easy for me to grab loads of information from the Internet, it's just as easy for other people to do the same. Who's out there searching me and what are they learning? And if Google has been watching every web search I've done, how much do they really know about me?

Fortunately, my anxiety was short-lived. Maybe it was the caffeine leaving my system or maybe the realization that a sensible person would focus their information-gathering efforts on rich celebrities and not an obscure statistician. In any case, my mind finally settled down, and I stopped asking questions and prepared to answer one. I finished grabbing the posting dates of all zombie flu-related hits that occurred between January 1 and May 1, 2009. My search definitely wasn't all-inclusive. I only searched Google and only looked for the term "Zombie Flu Outbreak," leaving out Twitter feeds and other social media as well as related terms. On the other hand, the Internet's largest search engine reaches far and wide, and these hits provide me with a good sample of data with which to work.

There's a little-known secret that professors like to keep under wraps and professional statisticians will only admit under duress. For a typical basic statistical analysis, a well-chosen plot does most of the work for you. Don't get me wrong, it's important to plug in the numbers, work the calculations, and look at the official results. But rarely does a formal analysis disagree with the intuition that comes from visualizing your data in the right way. This is why I love graphs.

When you're looking for mathematical relationships between two variables, a scatterplot works wonders. A *scatterplot* is a plot of two variables against one another, y versus x, and it's a great way to visualize relationships, get a sense for variation, and check for outliers. Figure 9.1 shows a scatterplot of zombie flu-related postings during the

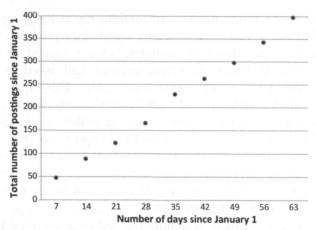

Figure 9.1.   Total number of zombie flu postings since January 1, 2009.

first two months of 2009, tallied by week. The x-axis plots days since January 1, and the y-axis plots the total number of postings in that time. These data provide me with a baseline, an indication of the zombie flu-related web chatter during January and February, before the topic went viral. The points on the scatter plot follow a clear pattern, with the value of y dependent on the value of x. If you imagine a line from the points in the lower right-hand corner to the upper-left, it looks as if most of the values would sit on the line or very close to it. These data clearly call for linear regression.

In linear regression, a dependent variable, y, and an independent variable, x, are related to one another with the function $y = mx + b$. The purpose of linear regression is to take a sample and estimate the slope, m, and the y-intercept term, b. As usual, the estimates, the values calculated from the data, are specially noted by placing a hat over the original (unknown) parameter values, in other words, $\hat{m}$ and $\hat{b}$. Basic linear regression uses a technique known as *least squares fitting* to estimate the slope and the y-intercept. Least squares fitting finds the values $\hat{m}$ and $\hat{b}$ that minimize the squared error, or total squared deviation between the model (estimated) y values and the observed y values. If $\widehat{y_k} = \hat{m}x_k + \hat{b}$ represents the regression estimate of y_k, the total squared error is:

$$\left(y_1 - \hat{y}_1\right)^2 + \left(y_2 - \hat{y}_2\right)^2 + \cdots + \left(y_N - \hat{y}_N\right)^2 .$$

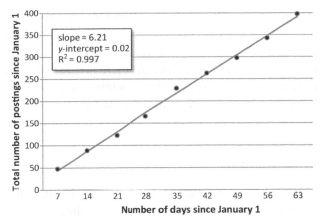

Figure 9.2. Regression analysis of zombie flu postings.

By making this function as small as possible, the regression procedure finds the parameters that, on the whole, make $\hat{y} = \hat{m}x + \hat{b}$ track the observed y-values as closely as possible. There are linear regression procedures in most basic data analysis software. For my purposes, I used the LINEST function in Excel. Figure 9.2 illustrates the results of linear regression performed on the zombie flu data for January and February 2009.

As indicated in the figure, the estimate $\hat{y}$ tracks the observed y-values closely and at first glance, it looks like the linear regression procedure worked well. The estimated slope $\hat{m} = 6.2$. Since the slope of a line measures the incremental change in y-values per unit change in x-values, this means there were, on average, about six new zombie flu-related posts per day during the first two months of 2009. The estimated intercept was $\hat{b} = 0.02$. The y-intercept of a line measures the y-value at $x = 0$. According to the regression line, then, there were basically no zombie flu postings between time $t = 0$ and the beginning of the year (which was time $t = 0$). This makes sense, and it should. No statistical analysis should ever contradict what you know to be true. If it does, then something's amiss.

A zero intercept is a common situation, and so many data analysis packages give you the option of automatically forcing the y-intercept $\hat{b}$ to be zero. I could have done this at the outset and then the estimated value of $\hat{b}$ would never have been a question. This is purely a matter of choice. I typically run the regression both ways, once with an estimated $\hat{b}$ term and once with it forced to zero, and then compare the results. The slope of the two lines should be almost the same, and the estimated intercept $\hat{b}$ should be nearly zero. If they're not, then either (a) your data do not fit a proper line, (b) there are outliers affecting the analysis, or (c) there's an error in your assumptions, your calculations, or both. In the case of the zombie flu data, the two slopes are both 6.21, and the estimated $\hat{b} = 0.02$ is so close to zero, it's not an issue.

Diagnosing the Fit

I now have a model for my data, in other words, a mathematical function for the number of zombie flu postings over time. But how good is it? Does the line pass through every one of the y-values? Is there a large error, or difference between the actual y-values and those predicted by the function I've just created? The plot in Figure 9.2 suggests the line is a good fit, and this is encouraging. There are also a number of *diagnostics*, tools for assessing the quality and accuracy of a model, that can be used to answer these questions.

R-Squared

Linear regression fits a line through a dataset. Statistically, this procedure measures something called correlation. Two random variables are *correlated* if the outcome of one affects the probability of the outcome of the other. For example, think of the zombie flu web postings data and pick a day at random. The probability of five or fewer postings up to this day depends on the day you choose. If you choose January 1, for example, the probability of five or fewer postings would be much higher than if you choose February 15. This means the number of days since January 1 and the total number of web postings since January 1 are correlated. (The concept of correlation may remind you of dependent random variables from Chapter 3. These two concepts are very similar, but not exactly the same.)

The *correlation coefficient* is a descriptive statistic that measures how highly correlated two variables are. The correlation coefficient is a value between −1 and 1. A value of −1 indicates the variables are inversely related to one another, and if you increase one, you'll see a decrease in the other. A value of 1 indicates the variables are positively related to one another, and if you increase one, you'll see an increase in the other. A value of 0 implies the two variables are uncorrelated, and the value of one does not impact the likely value of the other. For example, the correlation between the number of zombie flu postings and time is 0.999, suggesting the total number of posting is positively correlated with the number of days since January 1, 2009.

The correlation coefficient is one of the most common ways to evaluate the quality of a linear regression procedure, so common, in fact, it has its own name: R-squared. The R-squared, or R^2, value measures the correlation not between the x and y values but between the y and $\hat{y}$ values. If the regression line fits the data well, then y and $\hat{y}$ should track one another very closely and so R^2 should be close to one. If the line fits poorly or if excessive variation in y-values muddies the relationship between x and y, then R^2 will be closer to zero than one. Most linear regression programs calculate the R^2 as part of procedure. For the zombie flu regression in Figure 9.2, $R^2 = 0.997$.

The R^2 value is relatively easy to understand and so it's often the only diagnostic value people use to evaluate the quality of a regression line. This statistic isn't without its pitfalls, though. It works great when the range of x-values is big and there are no outliers impacting the analysis. But it can also be misleading. Figure 9.3 illustrates some different situations and how they impact the R^2 value. In Figure 9.3a–c, the R^2 value does a pretty good job of measuring how well the regression line fits the data. In Figure 9.3d, the line is a poor fit to the data, but a single outlier raises the R^2 value, suggesting the regression line fits better than it really does. Outliers like this one, on the edge of a regression region, are called *influential points*, because they can dramatically impact the quality of a regression fit. And since diagnostics like the R^2 statistics can also be impacted by influential points, it's important to look at your data and identify potential problems.

F-Test

In addition to the R^2 value, most linear regression procedures report the results of a hypothesis test for the slope. This test determines whether the slope of the regression line is significantly different from zero, meaning there really is a trend in the data. The hypotheses for this test are as follows:

H_0: $m = 0$

H_A: $m \neq 0$.

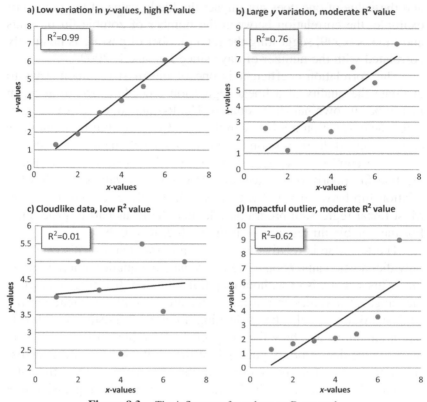

Figure 9.3. The influence of y-values on R-squared.

The hypotheses are similar to those for the t-test, in other words, the test for the mean of a population. However, the formula for the slope estimate $\hat{m}$ is such that the test statistic follows an F-distribution and not a t-distribution. The details of this test are straightforward but messy, and I refer you to the list of references at the back of the book for more details. For practical purposes, it's important to know that regression procedures report the results of an F-test for the slope of the regression line, and these results can be interpreted just like any other hypothesis test. The only real difference is in calculating the degrees of freedom. Recall, the F-distribution has two parameters: a numerator degrees of freedom and a denominator degrees of freedom. For simple linear regression with a slope and an intercept, there are two estimated parameters. The numerator degrees of freedom is the number of estimated parameters minus one, which is always one for simple linear regression. The denominator degrees of freedom is the number of observations in the sample minus the number of estimated parameters, or $N - 2$ for simple linear regression. These parameters can be used along with critical values for the F-distribution to determine the outcome of the test. For example, the F-statistic for the zombie flu regression is $F = 2035$. The $\alpha = 0.05$ value for the F-distribution with one numerator degree of freedom and $N - 2 = 7$ denominator degrees of freedom is $F_{crit} = 5.6$. Since $F = 2035$ is greater than 5.6, then we reject the null hypothesis, concluding that the slope of the line is statistically significant.

Residual Analysis

Correlation and hypothesis tests are helpful, but I find it even more helpful to visualize the results. That's why I'm a big fan of residual analysis. *Residuals* show you the statistical leftovers, the part of your data that isn't included in the regression line. Formally, the residuals are just the observations, the actual y-values, minus the corresponding estimated y-values, or $\hat{y}$. *Residual analysis* is the process of inspecting these residuals with a scatterplot. Figure 9.4 plots the residuals for the

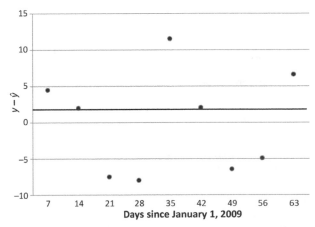

Figure 9.4.   Residuals of zombie flu regression line.

zombie flu data, the difference between the observed and the estimated number of postings during the first two months of 2009.

There are two things to look for in the residuals: outliers and non-linearity. As indicated in Figure 9.3, outliers can adversely impact the slope of the line, particularly if they lie at one end of range of x-values. There appear to be no obvious outliers in Figure 9.4 and so this probably isn't much of an issue for these data.

In a perfect world, the residuals should look completely random, without trends, without regular deviations from zero, and without runs of positive or negative values. In the real world, patterns sometimes appear. Patterns can be an indication of nonlinearity, meaning a linear regression line isn't enough to properly fit the data, or correlation, meaning the residuals are not independent of one another. Figure 9.5 shows patterns to look out for when examining residuals. When you see nonlinearity, you should try adding another x-variable or transforming the one you do have, and details on this can be found in a good textbook on linear regression. Correlation impacts the R^2, the tests

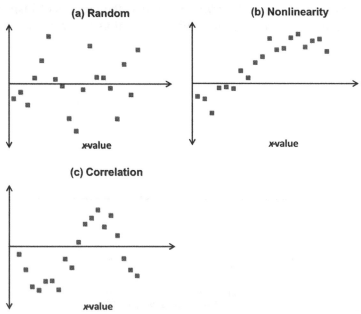

Figure 9.5. Things to look for in regression residuals.

for significance, prediction, and all other diagnostics that apply to a regression line. Mild correlation usually isn't a problem, but strong correlation can seriously mess with them, leading you astray on your predictions and your conclusions.

The residuals in Figure 9.4 do suggest a subtle pattern. They travel in twos: two above zero followed by two below zero and so on. This is an indication of weak correlation. The residuals are not independent of one another. However, this two-by-two pattern is obscured by the overall variation. So, it's probably safe to trust the diagnostics and move on.

Predicting the Spread of Infection

After all this analysis, I've fit a line to the zombie flu-related postings and convinced myself it's a good fit. But none of this answers my original question, "When did the zombie flu go viral?" It doesn't look like it happened in the first two months of 2009, since the related postings increased at a nice, steady, linear rate during that time. If not during January or February, when?

This is where prediction comes into play. *Prediction* is the ultimate goal of linear regression. It's where you take any x-value and use the fitted regression line to predict what the corresponding y-value should be. There are two kinds of prediction: *interpolation*, predicting for x-values inside the range of your fitted data, and *extrapolation*, extending the prediction beyond the x-range you originally used to fit the line. Interpolation is generally regarded as a trustworthy process. In contrast, most professors will tell you to be careful about extrapolating a regression function beyond the initial x-range of your data. This is good advice. If your goal is to predict y-values from x-values, then you should fit the regression line with a range of x-values surrounding those you're interested in. This gives you the most certainty in your predictions.

The goal of this study is not to predict y-values from x-values. The goal is simply to compare the predicted and actual number of zombie flu postings and determine when, if ever, the zombie flu went from a mere outbreak to a worldwide pandemic. In this case, extrapolation beyond the first two months of the year is exactly what I need. To do this, I need the *standard error of the regression line (SE)*. The SE measures the average deviation of the residuals, much in the way the standard deviation measures the average deviation of observations in a sample. The standard error estimates the variation of observed y-values around the regression line, and this value can be used to put a margin of error, or prediction interval, around a y-value. A prediction interval is similar to a confidence interval, but there's an important difference between the two. Confidence intervals apply to population parameters, things like the mean, and they measure how much uncertainty is associated with a parameter estimate. Prediction intervals apply to individual observations, and they measure how much uncertainty is associated with a single, estimated value, $\hat{y}$.

Constructing prediction intervals can be a little messy, and so I'll leave that to the more mathematical texts listed at the end of this book. But the general process is the same as a confidence interval: set up a $1 - \alpha$ interval probability around $\hat{y}$, then apply a sample distribution and some algebra to get the formula. For simple linear regression, the $(1 - \alpha)\%$ prediction interval for y is given by the following, rather unattractive formula

$$y = \hat{y} \pm SE \times t_{crit} \times \sqrt{1 + \frac{1}{N} + \frac{N(x_p - \overline{x})^2}{N(x_1^2 + \cdots + x_N^2) - (x_1 + \cdots + x_N)^2}}.$$

N is the number of observations used to fit the model, SE is the standard error of the regression line, t_{crit} is the $1 - \alpha/2$ critical value for the t-distribution with $N - 2$ degrees of freedom, x_p is the x-value for which you want the prediction, and $\overline{x}$ is the average of all the x-values used to fit the original line. For example, the zombie flu regression line for January and February has $N = 9$, SE $= 7.5$, and sample mean of x-values $\overline{x} = 35$. For a 95% prediction interval, t_{crit} is the 0.025 critical value for the t-distribution with $N - 2 = 7$ degrees of freedom. According to Appendix B, this value is $t_{crit} = 2.4$. The x-values used to fit the line were $x = 7, 14, 21, 28, 35, 42, 49, 56$, and 63. At $x_p = 70$ days, the estimated $\hat{y} = 434.9$. The 95% prediction interval for y is

$$y = 434.9 \pm 7.5 \times 2.4 \times \sqrt{1 + \frac{1}{9} + \frac{9(70 - 35)^2}{9(7^2 + \cdots + 63^2) - (7 + \cdots + 63)^2}}$$

$$= 434.9 \pm 26.22.$$

In other words, the true number of zombie flu postings in the first 70 days of 2009 was, in all likelihood, between 408.7 and 461.1.

Figure 9.6 extends the fitted regression line beyond the first two months of 2009, and plots the observed values on top of it. Figure 9.6a plots the total number of postings since January 1. Figure 9.6b gives us a closer look at the data by showing the residuals. The 95% prediction interval bounds are plotted alongside the residuals for reference. Notice how the bounds are tightest toward the middle of the fitted x-values, increasing on either side. This reflects the increasing uncertainty in the prediction as you move further from the center of the fitted x-range. It doesn't necessarily mean the estimate is less accurate. It just means you have a higher margin of error as you leave the safe range of fitted values.

The big jump in the total number of postings in April suggests the zombie flu did indeed go viral after the April Fool's Day prank news article; the number of posting jumped dramatically between April 20 and April 27. But it wasn't just the fake news article that did it. The idea had already started to spread before April 1. Figure 9.6b shows this. The actual number of postings left the range indicated by the 95% prediction interval by March 9, never to return. In other words, the disease was infecting the minds of Internet users for several weeks before the infamous prank. The April Fool's Day story might have been the trigger that made the zombie flu go viral, but these data suggest it was well on its way to urban myth before the article was ever posted.

FROM RUMOR TO MYTH TO LEGEND

What can we take away from this chapter? First, never give your bank account information to somebody claiming to be a rich person from another country. Rich people don't need your bank account. They have plenty of their own. Second, be careful what you do on the Internet. There are all sorts of people watching. I may be watching. And finally, if you're thinking of posting a story about a strange disease that turns your skin inside out or makes you develop a taste for human pancreas, you may not be the first. There are thousands of rumors, myths, and legends out there, circulating the Internet at lightning speed. There might already be a group of Internet junkies talking about this very condition, spreading the idea and infecting the minds of millions. But

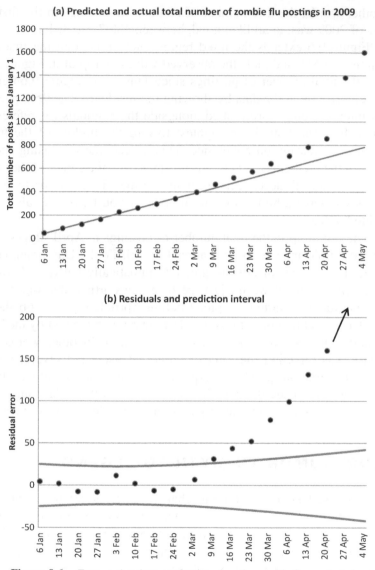

Figure 9.6. From outbreak to pandemic: when the zombie flu went viral.

even if your story isn't the first, it just might be the trigger, the one account that makes the disease go viral.

BIBLIOGRAPHY

BUTCHER, MIKE. "Tweetmeme Lets Hoax 'Zombie Swine Flu' BBC Story Go Unchecked." http://techcrunch.com/2009/05/01/london-is-not-quarantined-by-zombie-swine-flu-yet-tweetmeme-lets-hoax-bbc-story-go-unchecked/, accessed April 18, 2012.

SNOPES.COM. "Rumor Has It." www.snopes.com, accessed October 21, 2012.

WAYBACK MACHINE, INTERNET ARCHIVE. April 25, 2005. "BBC News, World Edition, Cambodian Troops Quarantine Quan'sul." http://web.archive.org/web/20050428004220/http://65.127.124.62/south_asia/4483241.stm.htm#top.

Critical Values for the Standard Normal Distribution

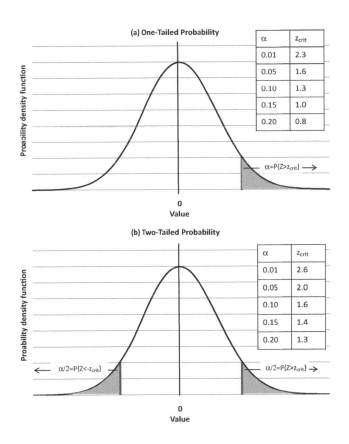

(a) One-Tailed Probability

α	z_{crit}
0.01	2.3
0.05	1.6
0.10	1.3
0.15	1.0
0.20	0.8

$\alpha = P\{Z > z_{crit}\} \longrightarrow$

(b) Two-Tailed Probability

α	z_{crit}
0.01	2.6
0.05	2.0
0.10	1.6
0.15	1.4
0.20	1.3

$\longleftarrow \alpha/2 = P\{Z < -z_{crit}\}$ $\alpha/2 = P\{Z > z_{crit}\} \longrightarrow$

The Art of Data Analysis: How to Answer Almost Any Question Using Basic Statistics,
First Edition. Kristin H. Jarman.
© 2013 John Wiley & Sons, Inc. Published 2013 by John Wiley & Sons, Inc.

Critical Values for the T-Distribution

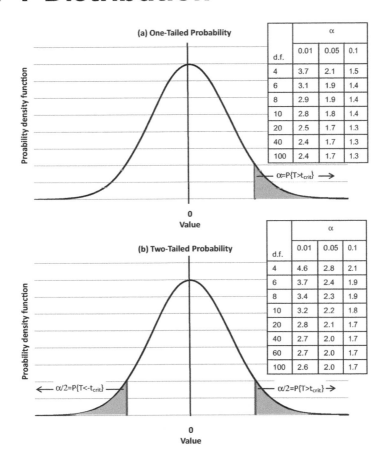

(a) One-Tailed Probability

d.f.	α		
	0.01	0.05	0.1
4	3.7	2.1	1.5
6	3.1	1.9	1.4
8	2.9	1.9	1.4
10	2.8	1.8	1.4
20	2.5	1.7	1.3
40	2.4	1.7	1.3
100	2.4	1.7	1.3

$\alpha=P\{T>t_{crit}\}$

(b) Two-Tailed Probability

d.f.	α		
	0.01	0.05	0.1
4	4.6	2.8	2.1
6	3.7	2.4	1.9
8	3.4	2.3	1.9
10	3.2	2.2	1.8
20	2.8	2.1	1.7
40	2.7	2.0	1.7
60	2.7	2.0	1.7
100	2.6	2.0	1.7

$\alpha/2=P\{T<-t_{crit}\}$ $\alpha/2=P\{T>t_{crit}\}$

The Art of Data Analysis: How to Answer Almost Any Question Using Basic Statistics,
First Edition. Kristin H. Jarman.
© 2013 John Wiley & Sons, Inc. Published 2013 by John Wiley & Sons, Inc.

APPENDIX C

Critical Values for the Chi-Squared Distribution

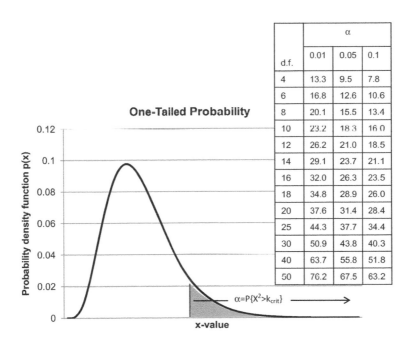

d.f.	α		
	0.01	0.05	0.1
4	13.3	9.5	7.8
6	16.8	12.6	10.6
8	20.1	15.5	13.4
10	23.2	18.3	16.0
12	26.2	21.0	18.5
14	29.1	23.7	21.1
16	32.0	26.3	23.5
18	34.8	28.9	26.0
20	37.6	31.4	28.4
25	44.3	37.7	34.4
30	50.9	43.8	40.3
40	63.7	55.8	51.8
50	76.2	67.5	63.2

The Art of Data Analysis: How to Answer Almost Any Question Using Basic Statistics,
First Edition. Kristin H. Jarman.
© 2013 John Wiley & Sons, Inc. Published 2013 by John Wiley & Sons, Inc.

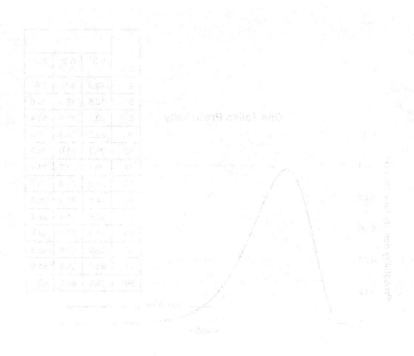

Glossary

Addition rule   For mutually exclusive events, the probability of their union is the sum of their probabilities.

Analysis of variance (ANOVA)   A hypothesis test comparing the mean of more than two groups or populations.

Average   A measure of the central value in a set of observations. Also called a sample mean. Calculate by adding all the values and dividing by the number of values you have.

Bernoulli distribution   A probability distribution describing a random variable with two possible outcomes and success probability p.

Binomial distribution   A probability distribution describing the number of successes in a fixed set of independent trials.

Blocking   In a study, the process of selecting members of a sample to be as homogeneous or similar as possible in order to minimize the impact of potentially confounding factors.

Classical probability   A probability calculated by dividing the number of outcomes in your event by the total number of outcomes in the sample space.

Complement   The complement of an event is the set of all outcomes not in that event.

Complement rule   For any event, the probability of its complement is one minus the probability of the event itself.

Conditional probability   The probability of A given B, or $P\{A|B\}$. The probability event A will occur given you know event B has occurred.

Confounding factor   A variable that affects the effect, or outcome, of an experiment, making your conclusions ambiguous.

The Art of Data Analysis: How to Answer Almost Any Question Using Basic Statistics,
First Edition. Kristin H. Jarman.
© 2013 John Wiley & Sons, Inc. Published 2013 by John Wiley & Sons, Inc.

Continuity correction   A 0.5 change to your x-value that adjusts for approximating a discrete distribution with a continuous distribution.

Controlled experiment   A data collection effort where variables or factors are tightly controlled.

Correlation   A relationship between two random variables or observations where the value of one affects the probability of the outcome of the other.

Counting rules   Formulas for counting the number of outcomes in an event or sample space.

Critical value   The x-value needed to achieve a specified probability.

Decision criterion   A rule for accepting or rejecting a hypothesis; usually a threshold above which the null hypothesis is rejected.

Degrees of freedom   The effective number of independent observations for a statistic, typically the number of observations minus one less the number of estimates needed.

Dependent variable   A variable whose value depends on some other variable.

Descriptive statistics   Values that summarize characteristics of a sample.

Diagnostics   Tools for assessing the quality and accuracy of a statistical model such as a regression line.

Effect   In a study, the outcome or phenomenon you'd like to measure.

Empirical probability   A probability value calculated from data and not just a theoretical model.

Empty set   An event with nothing in it. The intersection of two mutually exclusive events.

Estimate   A value calculated from a sample that estimates the corresponding value for an entire population.

Event   A specific set of outcomes in a random experiment.

Experimental design   The science of planning experiments to produce data that will lead to clear, valid conclusions.

Extrapolation   For a regression function, predicting y-values that lie outside the x-range of the original line.

Factor   Any variable that can impact the outcome of an experiment.

Frequency distribution   A list of numbers, one for each group or category in a qualitative dataset. The numbers count how many members of your dataset fall into each category.

Goodness-of-fit test   A hypothesis test that compares a frequency distribution to some model probability distribution, with the goal of judging whether or not the data fits the model.

Histogram   A method for graphing the frequency distribution of quantitative data. The data are binned, the number of observations in each bin are counted, and a bar graph of the counts is constructed.

Independent events   Two events where the probability of one occurring is unaffected by whether or not the other occurs.

Independent trials   Successive random experiments where the outcomes are independent.

Independent variable   A variable or factor that does not depend on any other variable or factor.

Influential points   Outliers or other values that have a dramatic effect on the results of least squares regression.

Interpolation   For a regression function, predicting y-values that lie inside the x-range of the original line.

Intersection   The set of all outcomes appearing in two or more events at once.

Interval probability   The probability that some random variable will fall within a range of values.

Least squares regression   The process of estimating a regression parameter by minimizing the squared error, or the total squared deviation between the estimated y values and the observed y values.

Linear regression   A method for relating two variables, x and y, through a linear function, namely, $y = mx + b$.

Median   A measure of the central value in a group of observations. It's the middle value.

Mode   The most popular category of a qualitative dataset.

Multiple comparison procedure   The process of testing for equality of means between many groups, two at a time.

Multiplication rule   For independent events, the probability of their intersection is the product of their probabilities.

Mutually exclusive events   Two events that cannot occur at the same time.

Normal distribution   A continuous probability distribution with a symmetric, bell-shaped curve.

Observational study   A data collection effort where none of the variables or factors are controlled.

Observations   Measurements, opinions, categories, or numerical values, anything that can make up a dataset.

Outliers   Extreme measurement values that can adversely impact the results of a statistical analysis.

Paired data   Observations collected in pairs that are not independent of one another.

Parameter estimate   A value that approximates the parameter of a probability distribution.

Prediction   In linear regression, the process of predicting y-values from corresponding x-values.

Probability   The mathematics of uncertainty and randomness; a fraction, a value between zero and one that measures the likelihood a given event will occur.

Probability density function   The probability formulation for a continuous random variable.

Probability distribution   A mathematical equivalent to a histogram. A function describing the shape, character, and relative likelihoods of data that conform to it.

Probability distribution function   The probability formulation for a discrete random variable.

Qualitative data or observations   Observations that describe a category or type, such as hair color. This type of measurement cannot be sorted into a meaningful numerical order.

Quantitative data or observations   Numerical observations that can be sorted into a meaningful order.

Random experiment   A situation or trial where the outcome is not known beforehand.

Random sampling   Choosing members of a dataset at random.

Random variable   The foundation of a probability distribution. An algebraic description of the outcome of a random experiment.

Randomization   In a study, the process of randomly ordering your data collection.

Range   The largest minus the smallest measurement in a dataset.

Regression   A method for predicting a variable, y, from another variable, x, through a mathematical function.

Relative frequency distribution   A frequency distribution where the counts in each category are represented as fractions of the whole or, alternatively, percentages.

Residual analysis   Inspection of residuals to identify outliers and determine how well a regression line fits the individual y-values.

Residuals   Statistical leftovers; in a regression, the error, or the actual y-values minus the corresponding y-values predicted by the regression line.

Sample space   The collection of all possible outcomes in a random experiment.

Sampling   The process of choosing which members of the population to include in a dataset for analysis.

Sampling with replacement   The process of successively drawing members of a population, replacing the members between draws.

Sampling without replacement   The process of successively drawing members of a population without replacing the ones you drew at each step.

Scatterplot   A plot of two variables against one another, a way to visualize relationships between two variables.

Standard deviation   The most common measure of variation, based on the mean squared deviation around the average.

Standard error   The standard deviation of an estimate, say, the sample mean.

Standard normal distribution   A special case of the normal distribution, where the mean is zero and the variance is one.

Statistically significant   A trend, pattern, or difference that is larger than you would expect based on random variation alone.

Test statistic   A value calculated from estimates of population parameters, and used to test hypotheses.

Type I error probability   For a hypothesis test, the probability of rejecting H_0 when H_0 is true.

Union   The union of two events is the set of all outcomes appearing in either one or both events.

Variation   Differences between the values of a dataset.

Venn diagram   A pictorial representation of outcomes and events in a sample space.

References

This book barely scratches the surface of the huge field of statistics. Here's a list of useful references for the curious reader. Some of these books are personal favorites. Some are recommendations from my colleagues. All are being used in classrooms across the United States to teach introductory probability, statistics, regression, and design of experiments.

Berry, Donald A., and Bernard W. Lindgren. 1996. *Statistics: Theory and Methods*, 2nd ed. Duxbury/Wadsworth.

DeGroot, Morris H., and Mark J. Schervish. 2002. *Probability and Statistics*, 3rd ed. Addison-Wesley.

DeVeaux, Richard D., and Paul F. Velleman. 2005. *Introductory Statistics*, 2nd ed. Addison Wesley, Boston, MA.

Freund, John, and Perles, Benjamin M. 2007. *Modern Elementary Statistics*, 12th ed. Pearson/Prentice Hall.

Hastings, Kevin J. 1997. *Probability and Statistics*. Addison-Wesley Longman.

Hoel, Paul G., Sidney C. Port, and Charles J. Stone. 1971. *Introduction to Statistical Theory*. Houghton Mifflin.

Hogg, Robert V., and Elliot A. Tanis. 2000. *Probability and Statistical Inference*. 6th ed. Prentice-Hall.

Larsen, Richard J., and Morris L. Marx. 2001. *An Introduction to Mathematical Statistics and Its Applications*. Prentice-Hall.

Mann, Prem S. 2010. *Introductory Statistics*, 7th ed. John Wiley & Sons.

Montgomery, D. C. 2012. *Design and Analysis of Experiments*, 8th ed. John Wiley & Sons.

Montgomery, Douglas C., and Elizabeth A. Peck. 2001. *Introduction to Linear Regression Analysis*, 3rd ed. John Wiley & Sons.

Moore, David S. 2003. *The Basic Practice of Statistics*, 3rd ed. W.H. Freeman, Hoboken, NJ.

The Art of Data Analysis: How to Answer Almost Any Question Using Basic Statistics, First Edition. Kristin H. Jarman.
© 2013 John Wiley & Sons, Inc. Published 2013 by John Wiley & Sons, Inc.

Moore, S. D., and G. P. McCabe. 2009. *Introduction to the Practice of Statistics*, 6th ed. Freeman Press, Dallas, TX.

Ross, Sheldon. 2010. *A First Course in Probability*, 8th ed. Pearson.

Ross, Sheldon M. 2010. *Introduction to Probability Models*, 10th ed. Elsevier.

Strait, Peggy Tang. 1989. *A First Course in Probability and Statistics with Applications*. Harcourt Brace Jovanovich.

Triola, Mario F. 2004. *Elementary Statistics using Excel*, 2nd ed. Pearson Education, Upper Saddle River, NJ.

Triola, Mario F. 2010. *Essentials of Statistics*, 4th ed. Pearson.

Triola, Mario F. 2011. *Elementary Statistics*, 11th ed. Pearson, Upper Saddle River, NJ.

Wackerly, D. D. 2008. *Mathematical Statistics with Applications*, 6th ed. Duxbury.

Weimer, Richard C. 1993. *Statistics*. Wm. C. Brown.

Index

Addition rule, 55, 167
American Community Survey, 65–83
Analysis of Variance (ANOVA),
    130–137, 167
  between group variation, 135
  F-test, 133–134
  results of, 137
  within group variation, 135
Arithmetic mean. *See* Mean
Average. *See* Sample, mean

Batman, 43–62
Bell-shaped distribution, 33. *See also*
    Normal distribution
Bernoulli distribution, 69
Bigfoot, 7–22
Bigfoot Field Researchers Organization,
    18
Binomial distribution, 69–72, 75, 167
  independent trials and, 70
  normal approximation to, 80, 99, 126
Binomial random variable, 72, 74
  independent trials and, 70
  mean, 75
  variance, 75
Blocking, 10, 16, 139, 167. *See also*
    Experimental design

CAPTCHA, 146
Causation versus association, 12
Census Bureau. *See* United States
    Census Bureau

Central Limit Theorem, 82–84, 95, 113,
    121, 124
Central location, 35
  mean, 35
  median, 36
  mode, 29
Chi-squared distribution, 117, 134, 165
Chi-squared goodness-of-fit test. *See*
    Goodness-of-fit test
Chi-squared statistic, 117, 127
Classical probability, 47, 51, 58, 76,
    167
Combination rule, 50, 51
Complement, 51–53
Complement rule, 73–74
Conditional probability, 56, 167
  of independent events, 56
  of mutually exclusive events, 56
Confidence interval, 91–101
  mean, known variance, 93
  mean, unknown variance, 97
  proportion, 99
Confounding factor, 10, 15, 19, 132,
    139, 167. *See* also Experimental
    design
Continuity correction, 80, 168
Continuous data, 32
Continuous probability distribution,
    75–81
Continuous random variable, 75
Controlled experiment, 10–12, 168
Correlation, 150, 168

The Art of Data Analysis: How to Answer Almost Any Question Using Basic Statistics,
First Edition. Kristin H. Jarman.
© 2013 John Wiley & Sons, Inc. Published 2013 by John Wiley & Sons, Inc.

Printed and bound by CPI Group (UK) Ltd, Croydon, CR0 4YY

27/10/2024

14580272-0002